KEYNOTE guide to topics in your course

Algae	page 65
Angiosperms	93
Bacteria	69
Bryophytes	77
Cells and Tissues	9
Chemistry	5
Diffusion	29
Digestion	21
Ecology	101
Evolution	97
Ferns	85
Flowers	41
Fruits	45
Fungi	73
Growth and Differentiation	57
Gymnosperms	89
Heredity	53
Leaves	13
Lycopsida	81
Meiosis and Nucleic Acids	49
Mineral Nutrition and Soils	33
Osmosis	29
Photosynthesis	17
Plant Kingdom	61
Psilopsida	81
Respiration	21
Roots	25
Seeds	45
Slime Molds	73
Sphenopsida	81
Stems	37
Tropisms	57
Viruses	69

CLIFFS KEYNOTE REVIEWS

Botany

by

JOAN E. RAHN, Ph.D.

CLIFF'S NOTES, INC. • LINCOLN, NEBRASKA 68501

ISBN 0-8220-1721-0

L. C. Catalogue Card Number: 70-89833

Printed in the United States of America

CONTENTS

		page
1	The Science of Botany	1
2	Some Basic Chemistry for Botany	5
3	The Plant Cell and Tissues	9
4	Leaves	13
5	Photosynthesis	17
6	Digestion and Respiration	21
7	Roots	25
8	Diffusion and Osmosis	29
9	Mineral Nutrition and Soils	33
10	Stems	37
11	Flowers	41
12	Seeds and Fruits	45
13	Meiosis and Nucleic Acids	49
14	Heredity	53
15	Growth, Differentiation, and Tropisms	57
16	Introduction to the Plant Kingdom	61
17	Algae	65
18	Bacteria and Viruses	69
19	Slime Molds and True Fungi	73
20	Bryophytes	77
21	Psilopsida, Lycopsida, and Sphenopsida	81
22	Ferns	85
23	Gymnosperms	89
24	Angiosperms	93
25	Evolution	97
26	Plant Ecology	101
	Final Examination	106
	Dictionary-Index	111

TO THE STUDENT

This KEYNOTE is a flexible study aid designed to help you REVIEW YOUR COURSE QUICKLY and USE YOUR TIME TO STUDY ONLY MATERIAL YOU DON'T KNOW.

FOR GENERAL REVIEW

Take the SELF-TEST on the first page of any topic and turn the page to check your answers.

Read the EXPLANATIONS of any questions you answered incorrectly.

If you are satisfied with your understanding of the material, move on to another topic.

If you feel that you need further review, read the column of BASIC FACTS. For a more detailed discussion of the material, read the column of ADDITIONAL INFORMATION.

FOR QUICK REVIEW

Read the column of BASIC FACTS for a rapid review of the essentials of a topic and then take the SELF-TEST if you want to test your understanding of the material.

ADDITIONAL HELP FOR EXAMS

Review terms in the DICTIONARY-INDEX.
Test yourself by taking the sample FINAL EXAM.

1 THE SCIENCE OF BOTANY

SELF-TEST

DIRECTIONS: Write your answers (a, b, c, or d) on the numbered lines to the right. To check your answers turn the page. Study the explanations to any questions you missed.

1 ______
2 ______
3 ______
4 ______
5 ______
6 ______
7 ______
8 ______

1. Characteristics generally exhibited by living things include
   a reproduction of their own kind and expenditure of energy
   b adaptation to the environment and a chemistry based on inorganic compounds
   c utilization of foods and production of more energy than is used
   d possession of cell walls and localized growth

2. The complex chemical compounds of which living things are composed
   a also occur in nonliving things but respond to different physical and chemical laws
   b also occur in nonliving things and respond to the same physical and chemical laws
   c do not occur in nonliving things but respond to the same physical and chemical laws
   d do not occur in nonliving things and do not respond to the same physical and chemical laws

3. One of the differences between most plants and most animals is that
   a plants produce carbon dioxide but animals do not
   b animals utilize foods but plants do not
   c animals need oxygen but plants do not
   d plants manufacture their own food but animals do not

4. Most plants differ from most animals in having
   a localized areas of growth and indefinite life spans
   b localized areas of growth and definite life spans
   c general growth throughout the body and indefinite life spans
   d general growth throughout the body and definite life spans

5. The vegetative organs of a seed plant are
   a leaf, stem, and seed
   b stamen, pistil, and seed
   c pistil, stamen, and root
   d root, leaf, and stem

6. As a general rule
   a branches, leaves, roots, and flowers are produced by stems
   b leaves, branches, and flowers are produced by stems, and branch roots arise from older roots
   c stems are produced by roots, and all other organs are produced by stems
   d leaves arise from internodes, and branches arise at nodes

7. Buds typically are found
   a at the tips of branches and along roots
   b at the tips of branches and at the bases of leaves
   c along roots and at the bases of leaves
   d only on the bases of leaves

8. A scientific theory is
   a an educated guess that a scientist plans to test by an experiment
   b a conclusion that a scientist reaches after performing a preliminary experiment
   c supported by the results of many related experiments but is still subject to revisions
   d supported by the results of so many experiments that it is not likely to be revised

BASIC FACTS

BOTANY is the science that deals with plants. It includes all aspects of their lives.

Being living things (organisms), PLANTS SHARE CERTAIN CHARACTERISTICS WITH ANIMALS. They GROW, increasing in size and weight. They have a complex CHEMICAL AND STRUCTURAL ORGANIZATION. The total of the chemical reactions occurring within organisms is called METABOLISM. Organisms have the property of IRRITABILITY, which is the ability to react to changes in their environment. They all exhibit some type of MOVEMENT, and they REPRODUCE their own kind. These processes involve the CONSTANT EXPENDITURE OF ENERGY.

PLANTS AND ANIMALS DIFFER in several ways. Most plants are fixed in one location, whereas most animals are capable of locomotion. Plants, however, are capable of GROWTH MOVEMENTS. Plants manufacture their own FOOD; animals do not have this ability and are dependent on plants for their food. GROWTH of plants is confined to certain localized areas called meristems, and it continues throughout the life of the plant. Growth generally occurs throughout the entire animal body until adult size is achieved, and then it stops. Plants usually grow indefinitely, but animals usually have definite life spans. Starch is the principal STORED FOOD of plants. Animals store food in their bodies in the form of glycogen or fat. Plant cells are surrounded by cellulose CELL WALLS; animal cells have no walls.

The VEGETATIVE ORGANS of a seed plant are leaves, root, and stems. The ROOT and its branches anchor the plant in the soil and absorb nearly all the water and minerals used by the plant. Many roots also store food. The LEAF is the primary food-synthesizing organ of seed plants. STEMS and their branches support leaves, flowers, fruits, and cones in the air;

(Continued on page 4)

ADDITIONAL INFORMATION

There is no single characteristic that can be used to define LIFE. Many of the characteristics of living things are shared by nonliving things, but only living things have all of the following traits: growth, complex organization, metabolism, irritability, movement, reproduction, and expenditure of energy.

In general the CHEMICAL COMPOUNDS of which organisms are composed are more complex than those of nonliving things. Except for viruses, the chemical constituents are organized into cells. In multicellular organisms the cells may be organized into tissues and tissues into organs.

METABOLISM includes anabolic and catabolic processes. ANABOLIC (synthetic) processes are those which form complex substances (foods) from simpler ones, while CATABOLIC (destructive) processes form simpler substances from complex ones.

The MOVEMENTS of organisms are not always obvious. Plant growth movements in response to stimuli such as light and temperature are slow. The contents of living cells show movement that is confined within the cells.

REPRODUCTION produces offspring which resemble their parents but which are not identical with them. Such VARIABILITY is due to mutations or to new combinations of characteristics.

No one characteristic or group of characteristics adequately differentiates PLANTS FROM ANIMALS. There are many exceptions to the characteristics which we associate with each group. Most of these exceptions are found among primitive plants and animals; indeed there is some debate whether certain protozoa (single-celled animals), algae, and fungi are plants or animals. Some protozoa are photosynthetic and a few have cell walls. Fungi are not photosynthetic, and some of them lack cell walls during part of their lives. Also, the response to stimuli is faster in some plants than in some animals.

Botany covers a large field and has a number of subdivisions. PLANT MORPHOLOGY is the study of the structure and form of plants. Included in morphology are PLANT ANATOMY, the study of plant tissues, and PLANT CYTOLOGY, the study of plant cells. PLANT PHYSIOLOGY deals with the functions of the parts of plants and includes PLANT BIOCHEMISTRY which concerns itself with chemical processes within plants. The naming, classification, and identification of plants is PLANT TAXONOMY. Related to both morphology and taxonomy are the studies of certain groups of plants such as algae (ALGOLOGY OR PHYCOLOGY), fungi (MYCOLOGY), bacteria (BACTERIOLOGY), and mosses and liverworts (BRYOLOGY). PLANT ECOLOGY determines the relationship of plants with their environments. Closely related to it is PLANT GEOGRAPHY which is concerned with the distribution of plants. PLANT GENETICS is the study of heredity. PALEOBOTANY is

(Continued on page 4)

EXPLANATIONS

1. *Characteristics generally exhibited by living things include* reproduction of their own kind, expenditure of energy (rather than its production), adaptation to the environment, and the utilization of foods. The chemistry of all living things is based on organic compounds. Possession of cell walls and localized growth are plant characteristics not shared by most animals.

2. *The complex chemical compounds of which living things are composed* do not occur in nonliving things but respond to the same physical and chemical laws. Living things possess no "Vital Principle" which was once thought to distinguish them from nonliving things. The chemical compounds in living things are usually more complex than those of nonliving things.

3. *One of the differences between most plants and most animals is that* plants manufacture their own food but animals do not. All living things utilize foods, and they all require oxygen in some form.

4. *Most plants differ from most animals in having* localized areas of growth and indefinite life spans. Most animals exhibit general growth throughout their bodies for a specific period of time. Growth then ceases. Each species of animal usually has a definite life span.

5. *The vegetative organs of a seed plant are* root, leaf, and stem. Stamens, pistils, and seeds are reproductive structures.

6. *As a general rule* leaves, branches, and flowers are produced by stems, and branch roots arise from older roots. Fruits and seeds are produced by flowers.

7. *Buds typically are found* at the tips of branches and at the bases of leaves. A few plants have the ability to produce buds on roots.

8. *A scientific theory is* supported by the results of many related experiments but is still subject to revisions. A law is supported by so much experimentation that it is not so likely to be revised, but in practice laws are not immutable.

Answers

a	1
c	2
d	3
a	4
d	5
b	6
b	7
c	8

they also transport food, water, and minerals between these organs and the root. A SHOOT is a stem of a young plant and its leaves. The term is also used to indicate a newly formed branch and its leaves. A NODE is the part of a stem that bears (or once bore) a leaf. The region of the stem between two successive nodes is called an INTERNODE. BUDS consist of several organs. Each bud contains a stem tip covered by bud scales, which are modified leaves. Young leaves, young flowers, or both leaves and flowers are produced on the stem tip. Branches ordinarily end in TERMINAL BUDS. LATERAL (or AXILLARY) BUDS arise at the nodes.

The REPRODUCTIVE ORGANS of a seed plant are either flowers, fruits, and seeds or cones and seeds. Each flower may produce one or more FRUITS. A fruit may contain one or more SEEDS, and CONES bear several seeds. In each seed is an EMBRYO, or young plant. A cluster of closely associated flowers is called an INFLORESCENCE.

As a science, botany lends itself to experimental procedure (SCIENTIFIC METHOD). This involves the following steps:

1. OBSERVATION and the raising of a QUESTION.
2. Construction of a HYPOTHESIS.
3. Testing of the hypothesis (EXPERIMENTATION).
4. Formulation of a THEORY.

an evolutionary field which deals with fossils and extinct plants. These subdivisions are all interdependent and related to each other. Specialization has advanced so far that some scientists are concerned with very narrow fields such as bacterial biochemistry or the cytology of a single group of algae.

In the first step of SCIENTIFIC METHOD, a QUESTION that is to be answered is formulated. This, of course, can be done only after the scientist has observed objects or phenomena and has become familiar enough with them to recognize that an unanswered problem exists. The question to be framed should be a simple one, for a question that is too broad cannot be easily answered in one experiment.

After making his HYPOTHESIS the experimenter turns to EXPERIMENT. No matter how reasonable the hypothesis sounds it cannot be accepted until it is supported by proof from experiment. The organisms used in an experiment must all be as nearly alike as possible. Not only should they be the same species, but also the same variety. They should also be of the same age and general history. If these precautions are not taken, the responses of the experimentals and controls may be due to their differences rather than to the differences in their treatments.

Every experiment should be repeated several times. If after repeated experiments the responses support the hypothesis, the hypothesis is accepted as valid. If the responses do not support the hypothesis, it must be rejected or amended. An amended hypothesis is then tested by further experimentation.

If an amended hypothesis is supported by additional experiments performed on many other organisms, it becomes a THEORY. Within limits a theory predicts what will happen under a certain set of circumstances. Theories are also subject to change as new knowledge becomes available. They may have to be revised or even rejected. A LAW is a theory that has been retested many times and found to have universal validity.

One exciting aspect of research is the fact that a completed experiment usually generates more questions. Experiments devised to answer these questions lead to still more questions. In this way our vast body of scientific information has accumulated and will continue to grow.

2 SOME BASIC CHEMISTRY FOR BOTANY

SELF-TEST

1. Isotopes of a given element differ in the number of
   a electrons
   b neutrons
   c protons
   d valence bonds

2. A chemical reaction involves the sharing of
   a electrons
   b protons
   c neutrons
   d ions

3. $\begin{array}{c}H\\ |\\ H—O—C—C—N—H\end{array}$ is unlikely to be the formula for a compound because
   a hydrogen has a valence of 2
   b carbon has a valence of 4
   c oxygen has a valence of 3
   d nitrogen has a valence of 4

4. The most common elements in plants are
   a oxygen, sulfur, carbon, and phosphorus
   b nitrogen, oxygen, phosphorus, and sulfur
   c sulfur, carbon, phosphorus, and nitrogen
   d carbon, hydrogen, oxygen, and nitrogen

5. The repeating unit in starch and cellulose molecules is
   a fructose
   b glucose
   c ribose
   d sucrose

6. The formula $C_6H_{12}O_6$ indicates a(n)
   a amino acid
   b fatty acid
   c monosaccharide
   d disaccharide

7. The formula for a fatty acid could be
   a $H_2C(NH_2)—C(=O)—OH$
   b $H_3C—CH_2—CH_2—C(=O)—OH$
   c $\begin{array}{ccccccc} & & H & & H & & H\\ & & | & & | & & |\\ H & — & C & — & C & — & C{=}O\\ & & | & & | & & \\ & & O & & O & & \\ & & | & & | & & \\ & & H & & H & & \end{array}$
   d $CH_3—CH_2OH$

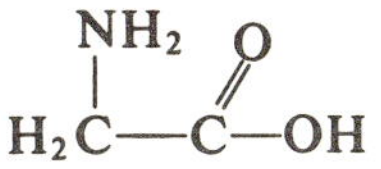

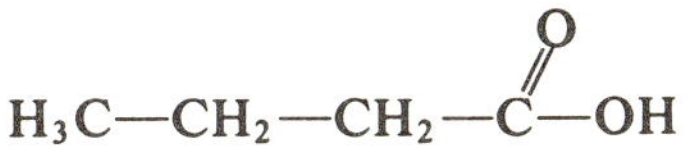

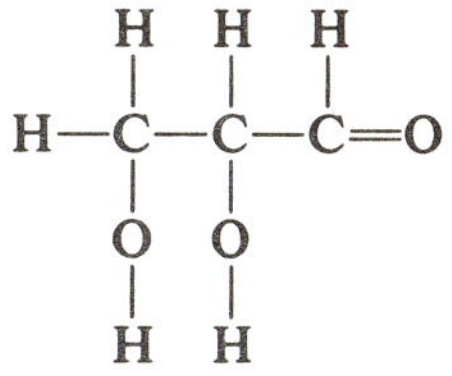

8. The lipids include
   a glucose, colloids, and fats
   b oils, glycerol, and colloids
   c fats, oils, and waxes
   d waxes, glycerol, and glucose

9. Amino acid molecules differ from sugar molecules in containing
   a nitrogen atoms
   b carbon atoms
   c phosphorus atoms
   d calcium atoms

10. Macromolecules common in plants are
    a nucleic acids, fatty acids, amino acids, and sucrose
    b nucleic acids, cellulose, starch, and maltose
    c glucose, sucrose, maltose, and ribose
    d starch, cellulose, proteins, and nucleic acids

11. Colloids are too small to be seen with a(n)
    a light microscope but are composed of macromolecules
    b light microscope and are composed of small molecules
    c electron microscope but are composed of macromolecules
    d electron microscope and are composed of small molecules

1 ______
2 ______
3 ______
4 ______
5 ______
6 ______
7 ______
8 ______
9 ______
10 ______
11 ______

BASIC FACTS

MATTER is anything that occupies space and has mass. It can neither be created nor destroyed in ordinary chemical reactions, but it is capable of undergoing change. ENERGY is the ability to do work. Energy is neither created nor destroyed in common chemical reactions, but one form of energy may be changed to another.

An ELEMENT is a substance that cannot be broken down into a simpler chemical substance by ordinary chemical means. An ATOM is the smallest unit of an element that can take part in a chemical reaction. CHEMICAL SYMBOLS are abbreviations for the names of elements and represent one atom of an element.

A COMPOUND is a substance composed of two or more elements chemically combined in definite proportions by weight. A MOLECULE is the smallest unit of a compound. A CHEMICAL FORMULA consists of the symbols of the elements of which a compound is composed and represents one molecule of the compound. A CHEMICAL EQUATION is shorthand for a chemical reaction. Formulas for the raw materials (reacting substances) are placed on the left side of the arrow; the formulas for the resulting products are placed on the right side.

ATOMS are composed of positively charged PROTONS, neutral NEUTRONS, and negatively charged ELECTRONS. Protons and neutrons constitute the nucleus of the atom; electrons circle the nucleus. The number of electrons of an atom is equal to the number of protons. However, some atoms of a particular element may have a different number of neutrons. These different types of atoms of an element are known as ISOTOPES.

In chemical reactions, atoms tend to gain or lose electrons or to share electrons with other atoms. The VALENCE of an element is the number of electrons its atoms tend to gain or lose (ELECTROVALENCE), or share (CO-

(Continued on page 8)

ADDITIONAL INFORMATION

The simplest of the CARBOHYDRATES are the MONOSACCHARIDES, or simple sugars, which have a ratio of one carbon atom to two hydrogen atoms to one oxygen atom. GLUCOSE and FRUCTOSE both have the formula $C_6H_{12}O_6$ and are called hexose sugars because they contain six carbon atoms. RIBOSE has the formula $C_5H_{10}O_5$ and is a pentose sugar. In monosaccharides most of the carbon atoms and one oxygen atom form a ring structure; the other atoms are attached variously to the carbons. Compounds like glucose and fructose that have the same molecular formulas but different structural arrangements are called ISOMERS.

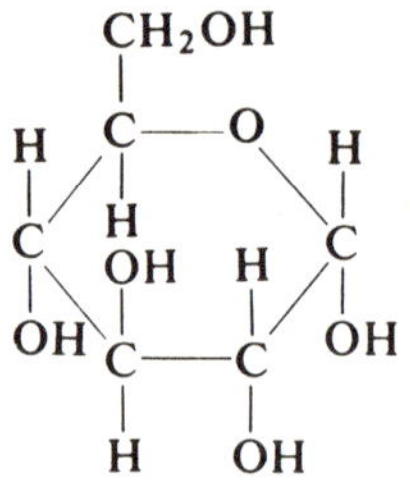

Glucose

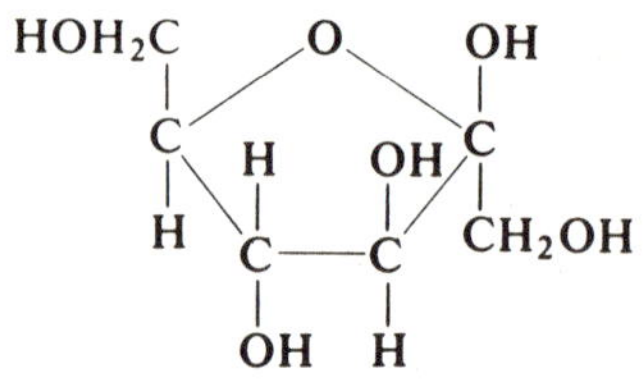

Fructose

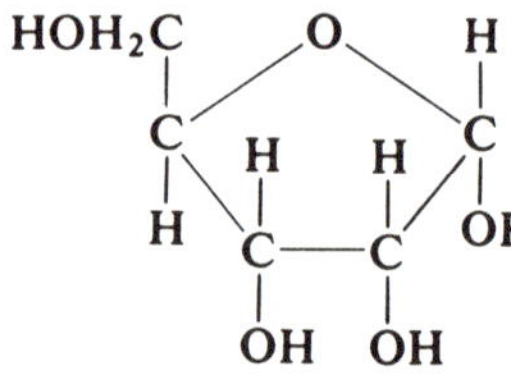

Ribose

A DISACCHARIDE is a sugar composed of two monosaccharides joined together with the loss of one molecule of water. Two disaccharides common in plants are MALTOSE, which is formed from two molecules of glucose, and SUCROSE, which is composed of one glucose and one fructose molecule.

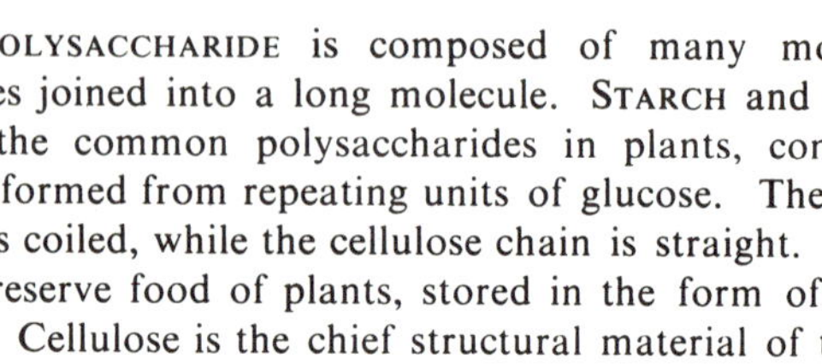

Sucrose

A POLYSACCHARIDE is composed of many monosaccharides joined into a long molecule. STARCH and CELLULOSE, the common polysaccharides in plants, consist of chains formed from repeating units of glucose. The starch chain is coiled, while the cellulose chain is straight. Starch is the reserve food of plants, stored in the form of starch grains. Cellulose is the chief structural material of the cell wall.

(Continued on page 8)

EXPLANATIONS

1. *Isotopes of a given element differ in the number of* neutrons in the nucleus of the atom. Valences and chemical properties, which depend on the number of electrons, are the same for all isotopes of a particular element.

2. *A chemical reaction involves the sharing of* electrons between atoms or the loss of an electron from one atom and its gain by another atom. A pair of shared electrons constitutes a chemical bond. Chemical reactions do not involve the particles of the atomic nucleus.

3. $H{-}O{-}C{-}C{-}\underset{}{\overset{\displaystyle H\atop|}{N}}{-}H$ *is unlikely to be the formula for a compound because* carbon has a valence of 4, not 2 as indicated in the structural formula. Hydrogen, oxygen, and nitrogen are correctly shown with valences of 1, 2, and 3, respectively.

4. *The most common elements in plants are* carbon, hydrogen, oxygen, and nitrogen. Nearly all organic compounds in plants contain carbon, hydrogen, and oxygen. Some of them contain nitrogen as well. Relatively few compounds in plants contain sulfur or phosphorus, but all of these elements are necessary for plants.

5. *The repeating unit in starch and cellulose molecules is* glucose. The glucose units are united in long chains.

6. *The formula* $C_6H_{12}O_6$ *indicates a* monosaccharide. Monosaccharides have a 1 to 2 to 1 ratio of carbon atoms to hydrogen atoms to oxygen atoms. Most of the monosaccharides in plants have three to seven carbon atoms per molecule.

7. *The formula for a fatty acid could be* $H_3C{-}CH_2{-}CH_2{-}\overset{\displaystyle O\atop /\!/}{C}{-}OH$. Fatty acids are straight-chain molecules composed of only carbon, hydrogen, and oxygen atoms. A fatty acid molecule terminates in an acid group: $-\overset{\displaystyle O\atop /\!/}{C}{-}OH$.

8. *The lipids include* fats, oils, and waxes; also fatty acids, phospholipids, and some other substances insoluble in water.

9. *Amino acid molecules differ from sugar molecules in containing* nitrogen atoms. Both amino acids and sugars contain carbon; neither contains phosphorus or calcium.

10. *Macromolecules common in plants are* starch, cellulose, proteins, and nucleic acids. These substances represent stored food, structural molecules, and regulatory substances. Macromolecules are large molecules composed of hundreds or thousands of atoms. Fatty acids, amino acids, glucose, sucrose, maltose, and ribose are relatively small organic molecules.

11. *Colloids are too small to be seen with a* light microscope but are composed of macromolecules. Proteins and polysaccharides are common colloidal substances in plants. Colloidal particles range in size from 0.001 μ to 0.1 μ. Particles of any one colloidal suspension remain in suspension primarily because they have similar electrical charges which repel each other.

Answers

b	1
a	2
b	3
d	4
b	5
c	6
b	7
c	8
a	9
d	10
a	11

VALENCE). In a STRUCTURAL FORMULA a chemical BOND (a pair of shared electrons) is indicated by a line:

$$\underset{\text{Water}}{H-O-H} \qquad \underset{\text{Carbon dioxide}}{O{=}C{=}O}$$

IONS are atoms or groups of atoms that have lost or gained electrons and thus bear positive or negative charges respectively.

CHEMICAL BONDS contain energy. Bonds between different kinds of atoms have different amounts of energy. When a chemical bond is broken, the bond energy is released.

ORGANIC COMPOUNDS are compounds containing carbon. Because of the ability of carbon atoms to form bonds with each other in chains and ring structures, many organic compounds are larger and more complex than inorganic compounds.

FOODS are organic compounds that can be used by organisms as a source of energy or as material for cell synthesis. The three main groups of foods are carbohydrates, fats, and proteins. CARBOHYDRATES contain the elements carbon, hydrogen, and oxygen. Hydrogen and oxygen are present in a ratio of 2 to 1. FATS contain the elements carbon, hydrogen, and oxygen, but the ratio of oxygen is much lower than it is in carbohydrates. Fats are richer in chemical energy than carbohydrates are. LIPIDS include the fats and fatlike substances that are insoluble in water. PROTEINS are large molecules containing the elements carbon, hydrogen, oxygen, and nitrogen, and usually sulfur.

NUCLEIC ACIDS (e.g., DNA, of which the genetic material is composed) are large organic molecules containing a pentose sugar, phosphate, and nitrogen-containing compounds called purines and pyrimidines.

COLLOIDS are MACROMOLECULES (large molecules) or clusters of molecules that are held in suspension in another substance. Macromolecules like proteins are often involved in colloids.

A FAT molecule is composed of one molecule of GLYCEROL and three molecules of FATTY ACIDS. Fatty acids are carbon chains of varying lengths. The end carbon always has two oxygen atoms attached to it: $-\overset{\overset{O}{\|}}{C}-OH$. If we let R stand for the variable portion of the molecule, then a general formula for a fatty acid is $R-\overset{\overset{O}{\|}}{C}-OH$.

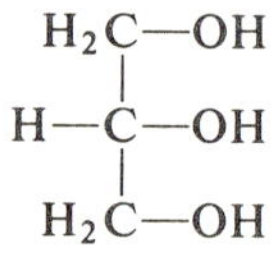

$$\begin{array}{l} H_2C-OH \\ \quad | \\ H-C-OH \\ \quad | \\ H_2C-OH \\ \text{Glycerol} \end{array} \qquad \begin{array}{l} H_2C-O-\overset{\overset{O}{\|}}{C}-R \\ \quad | \\ H-C-O-\overset{\overset{O}{\|}}{C}-R \\ \quad | \\ H_2C-O-\overset{\overset{O}{\|}}{C}-R \\ \text{A fat} \end{array}$$

The three fatty acids contributing to a fat need not be the same.

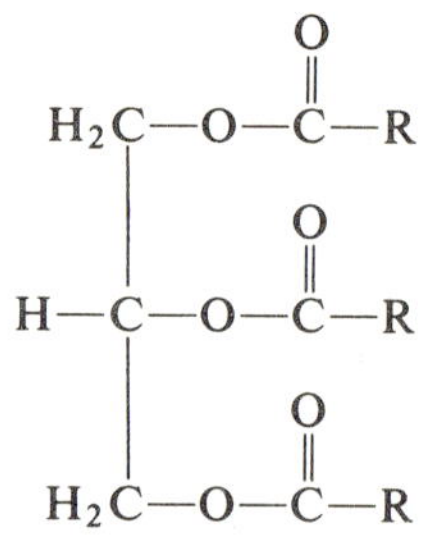

OILS are fats that are liquid at room temperature. Most plant fats are oils.

PHOSPHOLIPIDS are also formed from glycerol and fatty acids but contain phosphorus as well; they are important structural components of cell membranes. Other lipids important to the plant are WAXES. Cutin and suberin are composed of plant waxes.

PROTEINS are large molecules composed of AMINO ACIDS. Amino acids all contain carbon, hydrogen, oxygen, and nitrogen. Three of them contain sulfur. There are about 20 naturally occurring amino acids of varying composition, but they all have the following end group in common:

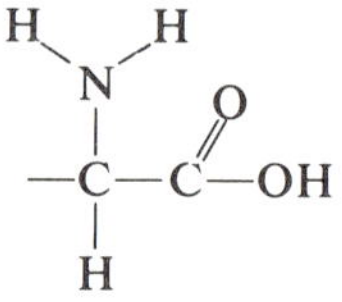

$$\begin{array}{c} H \quad H \\ \diagdown\;\diagup \\ N \qquad O \\ | \qquad /\!\!/ \\ -C-C-OH \\ | \qquad\quad \\ H \qquad\quad \end{array}$$

The $-COOH$ is the acid (carboxyl) group. The $-NH_2$ group is the amino group. If we let R stand for the variable portion of the molecule, then a general formula for an amino acid is:

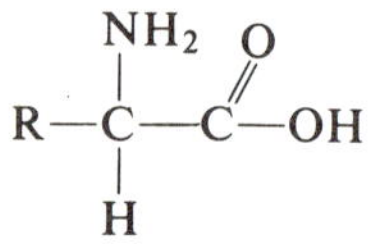

$$\begin{array}{c} NH_2 \quad O \\ | \qquad /\!\!/ \\ R-C-C-OH \\ | \qquad\quad \\ H \qquad\quad \end{array}$$

In the protein molecule, the amino acids are joined together through the amino group of one and the carboxyl group of another, with the loss of one molecule of water. This type of bonding is called the PEPTIDE LINKAGE, $-\overset{\overset{O}{\|}}{C}-\overset{\overset{H}{|}}{N}-$.

3 THE PLANT CELL AND TISSUES

SELF-TEST

1. The living portion of a plant cell is
   a cytoplasm
   b nucleus
   c protoplasm
   d cell wall

2. Membranes are found within
   a chromosomes, nuclei, and mitochondria
   b cytoplasm, chloroplasts, and mitochondria
   c cytoplasm, nuclei, and starch grains
   d chromosomes, chloroplasts, and starch grains

3. Membranes contain primarily
   a lipids and proteins
   b carbohydrates and lipids
   c nucleic acids and carbohydrates
   d nucleic acids and proteins

4. The function of chloroplasts is
   a respiration
   b digestion
   c photosynthesis
   d condensation

5. External protective tissues of plants are
   a epidermis and cork
   b cork and pericycle
   c cortex and epidermis
   d pericycle and cortex

6. Meristematic tissues of plants include
   a vascular cambium, cork cambium, and the tips of mature leaves
   b root and stem tips, vascular cambium, and cork cambium
   c mature fruits and root and stem tips
   d tips of mature leaves and mature fruits

7. Strengthening tissues of plants are
   a xylem, parenchyma, and collenchyma
   b parenchyma, sclerenchyma, and collenchyma
   c parenchyma, xylem, and epidermis
   d xylem, sclerenchyma, and collenchyma

8. Steps in mitosis are
   a interphase, prophase, metaphase, and anaphase
   b prophase, metaphase, anaphase, and telophase
   c metaphase, anaphase, telophase, and interphase
   d anaphase, telophase, interphase, and prophase

9. As a result of mitosis, daughter cells
   a receive identical chromosomes
   b receive nonidentical chromosomes
   c receive the same number of chloroplasts
   d receive the same number of mitochondria

10. Enzymes are important compounds in plants because they
    a are oxidized for their energy content
    b are structural components of cytoplasm
    c can be synthesized with the expenditure of very little energy
    d catalyze nearly all chemical reactions in the cells

11. An increase in temperature
    a increases the rate of a chemical reaction in cells
    b decreases the rate of a chemical reaction in cells
    c may either increase or decrease the rate of a chemical reaction in cells
    d usually has no effect on the rate of chemical reactions in cells

1 _____
2 _____
3 _____
4 _____
5 _____
6 _____
7 _____
8 _____
9 _____
10 _____
11 _____

BASIC FACTS

The CELL is the fundamental unit of living things. The plant cell consists of a living portion, the PROTOPLAST, surrounded by a nonliving CELL WALL. The living substance of the cell is also referred to as PROTOPLASM; it is divided into two regions, the NUCLEUS and the CYTOPLASM.

The CELL WALL, which is secreted by the protoplast, consists of a thin outer layer, the PRIMARY WALL, composed mainly of cellulose. The primary walls of two adjacent cells are cemented together by an intercellular substance, the MIDDLE LAMELLA. Fibrous and woody cells have a thickened inner layer, the SECONDARY WALL, which may contain substances such as lignin (in wood) and suberin (in cork).

The CYTOPLASM is bounded by a differentially permeable PLASMA MEMBRANE that controls the entrance and exit of materials into and out of the cell. Within the cytoplasm is the ENDOPLASMIC RETICULUM, which is a complex network of folded membranes. The cytoplasm also contains organelles (small bodies with specific functions). RIBOSOMES are located on the endoplasmic reticulum and are the site of protein synthesis. PLASTIDS are of three types: CHLOROPLASTS, which contain chlorophyll and produce all the food of the plant; CHROMOPLASTS, which are yellow, red, or brown and have no known function; and LEUCOPLASTS, which are colorless and store food in the form of starch or oil. MITOCHONDRIA are small granular or rod-shaped bodies in the cytoplasm that are the site of aerobic respiration. The GOLGI APPARATUS, or DICTYOSOME, consists of a stack of flat, membrane-bound sacs; it is thought to synthesize certain large carbohydrates.

The NUCLEUS is bounded by a NUCLEAR MEMBRANE. In the NUCLEAR SAP are at least one NUCLEOLUS and a number of threadlike bodies consist-

(Continued on page 12)

ADDITIONAL INFORMATION

Although a plant is divided into cells, and cells contain organelles, these parts are not discrete, isolated units. The nuclear membrane is connected with the endoplasmic reticulum, which in turn is continuous with the plasma membrane. The nuclear membrane contains PORES through which materials pass between nucleus and cytoplasm. Primary cell walls of living cells have small openings through which PLASMODESMATA (strands of cytoplasm) connect the protoplasts of adjacent cells. Secondary cell walls may have cavities called PITS. Often the pits of two adjacent cells are located directly opposite each other with only the middle lamella and the primary cell walls remaining between them.

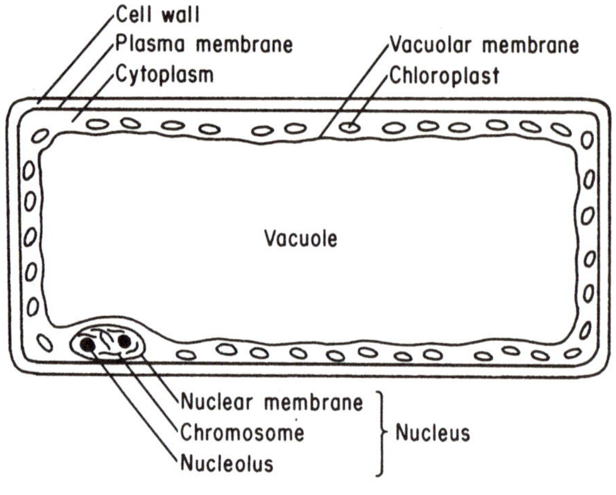

A Plant Cell

Growth involves the formation of new cells by the division of parent cells into daughter cells. Cell division involves MITOSIS, the division of the nucleus, and CYTOKINESIS, the division of the cytoplasm. The two processes usually occur simultaneously and are often referred to as mitosis. In INTERPHASE the nuclear membrane is intact with one or more nucleoli present in the nucleus. The chromatin threads are entangled and appear to form a net or reticulum. Shortly before prophase each chromosome duplicates itself. The two parts, called CHROMATIDS, are attached to each other at a point called the CENTROMERE (or KINETOCHORE). In PROPHASE the chromosomes become shorter and thicker, and the two chromatids of each chromosome become visible. The nucleoli and the nuclear membrane disappear. Late in prophase SPINDLE FIBERS form. They extend from two points (poles) at opposite ends of the cell. In METAPHASE the short, thick chromosomes move to the middle of the spindle. An additional fiber arises from the centromere of each chromatid and extends to one pole of the cell. In ANAPHASE the two chromatids of each chromosome separate and move to oppo-

(Continued on page 12)

EXPLANATIONS

1. *The living portion of a plant cell is* protoplasm. Protoplasm is divided into two parts, nucleus and cytoplasm, each of which contains smaller units in which vital chemical reactions occur. The cell wall is not living.

2. *Membranes are found within* cytoplasm, chloroplasts, and mitochondria. The endoplasmic reticulum of the cytoplasm is a membrane system. Both chloroplasts and mitochondria are bound by a double membrane. Chloroplasts contain small bodies called grana that are composed of membranous, disc-shaped lamellae stacked like a pile of coins; the grana are connected with each other by other membranes called frets, intergrana, or intergranum lamellae. The inner membrane of a mitochondrion is thrown into folds called cristae. Many important chemical reactions such as enzyme synthesis, photosynthesis, and respiration occur on or in association with membranes.

3. *Membranes contain primarily* lipids (fats and fatty substances) and proteins. Most cell membranes consist of two layers of proteins between which are two layers of lipids. Such membranes are strong, yet flexible. Some, if not all, of them contain enzymes, which are proteins.

4. *The function of the chloroplasts is* photosynthesis, a process which produces the food of the plant. Chloroplasts are the only structures within the cell that contain the photosynthetic pigment chlorophyll.

5. *External protective tissues of plants are* epidermis and cork which provide protection from excessive water loss and from invasion by parasites. Epidermis covers leaves and young roots and stems. Cork covers older roots and stems. Pericycle is an internal tissue of roots and underground stems.

6. *Meristematic tissues of plants include* root and stem tips, vascular cambium, and cork cambium. Meristems are embryonic tissues that provide new growth. Roots and stems grow in length at their tips. Vascular cambium and cork cambium provide growth in width of roots and stems. Young leaves and fruits are composed almost entirely of meristematic tissue, but growth ceases at maturity.

7. *Strengthening tissues of plants are* xylem, sclerenchyma, and collenchyma. Most strengthening tissues consist of elongated cells with thick cell walls. Parenchyma cells are usually thin-walled.

8. *Steps in mitosis are* prophase, metaphase, anaphase, and telophase. Nondividing nuclei are in interphase.

9. *As a result of mitosis, daughter cells* receive identical chromosomes. The distribution of cytoplasmic constituents to the daughter cells is rarely equal, although it may be nearly so. In mitosis each chromosome duplicates itself, and each daughter cell receives an identical set of chromosomes.

10. *Enzymes are important compounds in plants because they* catalyze nearly all chemical reactions in the cells. Catalysts are substances that increase the rates of chemical reactions. Enzymes are a special class of catalysts that are produced only by living cells.

11. *An increase in temperature* may either increase or decrease the rate of a chemical reaction in cells. Within a certain temperature range an increase in temperature increases the rate of a chemical reaction, but above a certain temperature the enzymes are denatured and can no longer speed reactions. The optimum temperature is different for different chemical reactions.

Answers

c	1
b	2
a	3
c	4
a	5
b	6
d	7
b	8
a	9
d	10
c	11

ing of CHROMATIN. When a cell divides (MITOSIS) the chromatin threads condense into rodlike bodies (CHROMOSOMES). One-half of each chromosome goes to each daughter cell. Thus each new cell receives chromosomes identical with those of the parent cell. The stages of mitosis are PROPHASE, METAPHASE, ANAPHASE, and TELOPHASE. A nucleus that is not dividing is in INTERPHASE.

VACUOLES, CRYSTALS, and STARCH GRAINS are nonliving inclusions in the cytoplasm. VACUOLES are droplets of water containing dissolved substances; they are separated from the cytoplasm by a VACUOLAR MEMBRANE. In mature cells the smaller vacuoles have coalesced into a CENTRAL VACUOLE that displaces the cytoplasm to a thin layer against the cell wall.

TISSUES are organized groups of cells having a specific function. PARENCHYMA is a relatively unspecialized, thin-walled tissue that stores food; it may also contain chloroplasts. SCLERENCHYMA and COLLENCHYMA consist of thick-walled cells that contribute support to the plant. Mature sclerenchyma cells are dead, but collenchyma cells are alive. The PROTECTIVE TISSUES are the EPIDERMIS and CORK. They reduce water loss and mechanical damage. The epidermis, usually one cell thick, has a waxy cover of cutin; it is a living tissue and is found in young leaves, stems, and roots. Cork is a dead tissue and occurs in old organs. The CONDUCTING, or VASCULAR, TISSUES of the plant are the xylem and phloem. XYLEM transports water and dissolved minerals upward in the plant and contributes to its support. PHLOEM transports food material both upward and downward. MERISTEMS are embryonic tissues that produce new cells. APICAL MERISTEMS occur at root and stem tips and provide increase in length of these organs.

site poles of the cell. Each chromatid now is called a chromosome. The sets of chromosomes moving to opposite poles are identical with each other. In TELOPHASE events occur that are the reverse of those of prophase. A nuclear membrane forms around each set of chromosomes. Nucleoli appear in the new nuclei, and the chromosomes become long and thin again. The nucleus now enters interphase. CYTOKINESIS occurs during telophase. A thin membrane, the CELL PLATE, forms across the center of the cell. The spindle begins to shrink toward the cell plate and disappears. Division of the cell into two parts has been completed. The cell plate remains as the middle lamella, and a primary cell wall forms on either side of it.

SIMPLE TISSUES are composed of only one type of cell. Xylem and phloem are COMPLEX TISSUES. Four kinds of cells are found in XYLEM: tracheids, vessels, fibers, and parenchyma. TRACHEIDS are elongated, lignified, pitted cells that serve for both conduction and support. VESSELS, which are open, continuous tubes, are tracheid-type cells modified for conduction. FIBERS are longer than tracheids and have thicker walls; they serve primarily for support. PHLOEM contains sieve tubes, companion cells, fibers, and parenchyma. SIEVE TUBES are the conducting cells; they are elongated with perforated end walls (SIEVE. PLATES) through which the cytoplasm of adjacent sieve elements is connected. COMPANION CELLS, which are adjacent to the sieve tubes, have nuclei; sieve tubes do not.

WATER is the most abundant INORGANIC COMPOUND in the plant cell, making up more than 90 percent of its living material. Mineral SALTS are also present, largely in ionic form; they are used by the plant in synthesizing organic compounds. The main groups of ORGANIC COMPOUNDS synthesized by plants and also constituting most of the dry weight of the cell are CARBOHYDRATES, LIPIDS, and PROTEINS. NUCLEOPROTEINS are proteins combined with nucleic acids; they are the chief constituent of the genes. ENZYMES, which catalyze nearly all the chemical reactions in cells, are complex proteins, some with nonprotein components. Without enzymes, cellular reactions would proceed too slowly for the maintenance of life. Enzymes are destroyed by high temperatures and their action is inhibited by certain chemicals (poisons). They are highly specific, each one catalyzing a specific reaction of a specific substance. Perhaps a thousand or more different enzymes are necessary for the normal functioning of a single cell.

4 LEAVES

SELF-TEST

1. The main functions of leaves are
   a photosynthesis and transpiration
   b transpiration and respiration
   c respiration and digestion
   d photosynthesis and respiration

2. The main parts of a leaf are
   a petiole and rachis
   b rachis and stipule
   c stipule and blade
   d blade and petiole

3. The typical monocot leaf differs from the typical dicot leaf in having
   a parallel venation and a sheathing leaf base
   b parallel venation and a petiole
   c net venation and a sheathing leaf base
   d net venation and a petiole

4. The leaf shown in the illustration is
   a palmately compound
   b palmately lobed
   c pinnately compound
   d pinnately lobed

5. The correct order of tissues in a typical leaf (beginning with the upper surface and proceeding to the lower) is
   a upper epidermis, spongy parenchyma cambium, palisade parenchyma, lower epidermis
   b upper epidermis, palisade parenchyma, cambium, spongy parenchyma, lower epidermis
   c upper epidermis, spongy parenchyma, palisade parenchyma, lower epidermis
   d upper epidermis, palisade parenchyma, spongy parenchyma, lower epidermis

6. In a typical leaf the highest concentration of chloroplasts is found in the
   a epidermis
   b spongy parenchyma
   c palisade parenchyma
   d vascular tissue

7. Stomata are located in
   a the upper epidermis only
   b the lower epidermis only
   c either the upper or lower epidermis, depending on the species
   d either the upper or lower epidermis or both, depending on the species

8. Thick cuticles on leaves are typical of plants growing in
   a wet habitats
   b dry habitats
   c warm habitats
   d cool habitats

9. Modified leaves may take the form of
   a stolons, tendrils, or cotyledons
   b spines, traps, or cotyledons
   c spines, traps, or corms
   d corms, stolons, or tendrils

10. Autumn coloration of leaves is due to the presence of
    a carotene, xanthophylls, and anthocyanins
    b chlorophyll, carotene, and xanthophylls
    c xanthophylls, anthocyanins, and phytochrome
    d phytochrome, carotene, and anthocyanins

11. Water and minerals are carried to the leaf from the roots by the
    a cambium
    b phloem
    c xylem
    d pith

1 ______
2 ______
3 ______
4 ______
5 ______
6 ______
7 ______
8 ______
9 ______
10 ______
11 ______

BASIC FACTS

The leaf is the main photosynthetic organ of the plant. It typically consists of a BLADE (the flat portion of the leaf) and a PETIOLE (a stalk which attaches the blade to the stem). SESSILE leaves lack petioles; the blades arise directly from the stem. In most monocots, SHEATHING LEAF BASES replace petioles. Some dicot leaves have two STIPULES (small appendages arising from either side of the lower portion of the petiole) which appear as small blades, thorns, or tendrils.

The petiole carries vascular bundles, or VEINS, that continue to run through the blade. Monocot leaves have PARALLEL VENATION with several veins of equal rank running approximately parallel to each other. Dicot leaves have NET VENATION. One main vein, the MIDRIB, branches into smaller veins which branch several more times, the smaller branches meeting each other and forming a network. In a PALMATELY VEINED leaf several major veins radiate from the apex of the petiole. In a PINNATELY VEINED leaf secondary veins branch off from the midrib at regular intervals.

Leaves are either simple or compound. The blade of a SIMPLE leaf is all in one piece; that of a COMPOUND leaf is divided into several leaflets. In a PALMATELY COMPOUND leaf all the leaflets arise from the same point at the apex of the petiole. In a PINNATELY COMPOUND leaf the leaflets arise along the length of the RACHIS, which is a continuation of the petiole.

The MARGINS of the leaf may be ENTIRE (SMOOTH), TOOTHED (SERRATE), or LOBED. In a PALMATELY LOBED leaf the lobes appear to radiate from the tip of the petiole; in a PINNATELY LOBED leaf the lobes are arranged along the length of the midrib.

The ARRANGEMENT of leaves on a stem may be ALTERNATE, OPPOSITE, or WHORLED.

(Continued on page 16)

ADDITIONAL INFORMATION

The CUTICLE is secreted by epidermal cells. It is usually thicker on the upper epidermal layer than on the lower. The STOMATA usually are more numerous in the lower epidermis than in the upper. When the GUARD CELLS of the stomata are turgid, or swollen with water, they separate; carbon dioxide from the air diffuses through the stomata into the leaf, and oxygen, produced during photosynthesis, diffuses out. When the guard cells are relaxed, they come together; the stomata are then closed. Generally, the stomata are open during the day, when photosynthesis takes place, and closed at night. TRANSPIRATION, the loss of water vapor, occurs primarily through the surfaces of the leaf. The cuticle and the closing of the stomata at night reduce the loss of water to a minimum. Plants living in arid climates usually have thicker cuticles on their leaves than do those living in aquatic environments.

The PALISADE PARENCHYMA contains more chloroplasts and produces more food than the SPONGY LAYER. The cells are close to each other. The large intercellular spaces of the spongy parenchyma permit carbon dioxide that enters through the stomata to diffuse laterally throughout the leaf and to reach all palisade cells.

Xylem occupies the upper portion of each VASCULAR BUNDLE and is continuous with the xylem of the stem. Phloem occupies the lower portion of each vascular bundle and is continuous with the phloem of the stem. Most vascular bundles are surrounded by a BUNDLE SHEATH of parenchyma cells. The bundle sheath increases the area of contact between the conducting tissues and the mesophyll cells. In some plants, bundle sheath extensions composed of collenchyma or sclerenchyma extend upward and downward to the epidermal layers. The vascular bundles, their sheaths, and their associated collenchyma or sclerenchyma contribute to the support of the leaf blade.

AQUATIC PLANTS grow with their roots and a considerable portion of their stems submerged in water. These organs require more oxygen than water can supply. Leaves and other organs of aquatic plants have extremely large intercellular spaces that provide a passageway by which oxygen can diffuse to the submerged portions.

In plants with LEAVES THAT STAND VERTICALLY, or nearly so, the distinction between palisade and spongy parenchymas is frequently lost. There is simply a single mesophyll tissue. Vertical leaves of grasses, for example, have no "top" or "bottom" surface with reference to the sun. During the course of the day both surfaces are likely to receive about the same amount of light, and differentiation into palisade and spongy parenchymas would have no advantage.

(Continued on page 16)

EXPLANATIONS

1. *The main functions of leaves are* photosynthesis (the manufacture of food) and transpiration (the loss of water in vapor form). Respiration and digestion are functions of cells in all organs of the plant.

2. *The main parts of a leaf are* considered to be the blade and the petiole even though some leaves lack petioles. Stipules are absent from a considerable number of leaves, and rachises are found only in pinnately compound leaves.

3. *The typical monocot leaf differs from the typical dicot leaf in having* parallel venation and a sheathing leaf base rather than a petiole. The leaf bases clasp the stem and may completely encircle it. Bases of successive leaves may overlap each other so as to cover the stem completely.

4. *The leaf shown in the illustration is* palmately lobed. The lobes radiate from the tip of the petiole.

5. *The correct order of tissues in a typical leaf (beginning with the upper surface and proceeding to the lower) is* upper epidermis, palisade parenchyma, spongy parenchyma, lower epidermis. Cambium is not found in leaves; it is usually confined to those organs that live for more than one year and that exhibit additional growth each year. Most leaves are retained for only one growing season. Even leaves of evergreen plants usually lack cambium.

6. *In a typical leaf the highest concentration of chloroplasts is found in the* palisade parenchyma. No vascular tissue and no epidermal cells except the guard cells of the stomata contain chloroplasts. The palisade layer receives more light than the spongy parenchyma, for sunlight passes through only the clear epidermal layer before reaching the palisade cells.

7. *Stomata are located in* either the upper or lower epidermis or both, depending on the species. Most species have more stomata in the lower epidermis than in the upper.

8. *Thick cuticles on leaves are typical of plants growing in* dry habitats. A thick cuticle reduces the amount of water lost in transpiration. Dry air has a low relative humidity which increases the rate of transpiration.

9. *Modified leaves may take the form of* spines, insect-catching traps, food-storage cotyledons of embryos, or tendrils. Modified leaves may or may not be photosynthetic.

10. *Autumn coloration of leaves is due to the presence of* one or more of the pigments carotene, xanthophylls, and anthocyanins. Chlorophyll disintegrates as autumn coloration develops, and phytochrome is present in such low concentrations as to contribute no color.

11. *Water and minerals are carried to the leaf from the roots by the* xylem; phloem removes manufactured food from the leaf.

Answers

a	1
d	2
a	3
b	4
d	5
c	6
d	7
b	8
b	9
a	10
c	11

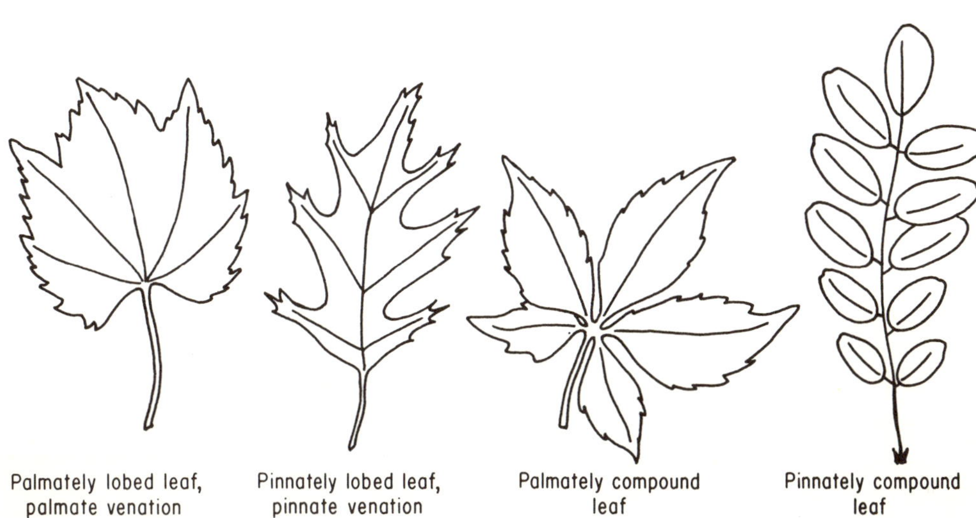

Palmately lobed leaf, palmate venation — Pinnately lobed leaf, pinnate venation — Palmately compound leaf — Pinnately compound leaf

Leaf Types

The leaf blade consists of three main types of TISSUE: the upper and lower epidermis, the mesophyll, and the vascular bundles. The EPIDERMIS protects the leaf from injury and desiccation. It is often covered by a waxy CUTICLE. Small openings in the epidermis, called STOMATA, permit gaseous exchange between the atmosphere and the mesophyll. Each stoma is surrounded by a pair of GUARD CELLS which regulate the size of the opening.

The MESOPHYLL is the main site of the CHLOROPLASTS, small bodies that contain the chlorophyll. The mesophyll has an upper layer, the PALISADE PARENCHYMA, which consists of one to three tiers of vertically elongated cells, and a lower layer, the SPONGY PARENCHYMA, the cells of which are widely spaced and of irregular shape. Carbon dioxide from the atmosphere entering through the stomata reaches the air spaces in the spongy parenchyma and diffuses throughout the leaf.

Each VASCULAR BUNDLE is composed of two tissues: the XYLEM, which conducts water and minerals from the roots through the stem to the leaf, and the PHLOEM, which conducts food manufactured in the mesophyll to the stem and from there to actively growing portions of the plant or to storage organs.

EVERGREENS retain their leaves throughout the year, individual leaves remaining on the plant for one or more years. New leaves are produced each spring. DECIDUOUS plants lose their leaves every autumn, and new leaves emerge from buds every spring. Late in summer an ABSCISSION ZONE develops across the petiole near its base. The layer is composed of thin-walled cells that gradually disintegrate.

In addition to FOLIAGE LEAVES, which are primarily concerned with photosynthesis, there are SPECIALIZED (MODIFIED) LEAVES that have many different functions. FOOD-STORAGE LEAVES include BULB SCALES, such as those of an onion, and COTYLEDONS, which are the first-formed leaves of the embryos of seed plants. WATER-STORAGE LEAVES, such as the thick, succulent leaves of some desert plants, have large parenchyma cells that form a water-storage tissue.

PROTECTIVE LEAVES include bud scales and spines. BUD SCALES, which are the outer leaves of buds, are thick, tough, and frequently resinous; they protect the tender growing points from desiccation, mechanical injury, and parasites. SPINES, such as those of the cactus, discourage grazing animals.

SUPPORTING LEAVES, or TENDRILS, are modified leaves or leaflets of climbing plants.

In certain plants, leaves have a REPRODUCTIVE function. Young plantlets normally develop in the marginal notches of *Bryophyllum leaves*. The leaves of African violets can be induced to form new plants by placing them in water.

Pitcher plants, sundews, and Venus's-flytraps have leaves modified for INSECT-TRAPPING.

Leaves of deciduous trees and shrubs often show AUTUMN COLORATION. Associated with chlorophyll in the chloroplasts are yellow to orange pigments, CAROTENE and XANTHOPHYLLS. In autumn, chlorophyll molecules decompose, revealing the orange and yellow pigments. If days have been sunny and nights have been cool, an excess of glucose accumulates in the leaf; this is used in the synthesis of ANTHOCYANINS, which are red or purplish pigments. The many shades of autumn coloration are produced by varying concentrations of these pigments.

5 PHOTOSYNTHESIS

SELF-TEST

1. The rate of photosynthesis increases when
   a the carbon dioxide concentration becomes extremely high
   b the light intensity becomes extremely high
   c the amount of water in the soil is increased by a heavy rainfall
   d the temperature is increased to a certain limit

2. In photosynthesis, light
   a is converted into kinetic energy
   b is converted into chemical energy
   c acts like a catalyst and remains unchanged
   d acts directly on carbon dioxide and water

3. Before light can be used in photosynthesis it must be
   a converted into heat
   b filtered to remove harmful wavelengths
   c absorbed by a pigment
   d absorbed by carbon dioxide

4. Light that is most effective in photosynthesis in green plants is
   a green
   b orange and yellow
   c red, blue, and violet
   d green, orange, and blue

5. When a compound gains electrons or hydrogen atoms it is said to be
   a hydroxylated
   b oxidized
   c reduced
   d hydrolyzed

6. The role of water in photosynthesis is
   a supplying minerals in solution
   b providing hydrogen with which carbon dioxide is reduced
   c providing oxygen to oxidize carbon dioxide
   d providing a solution in which chlorophyll can act

7. The steps in photosynthesis may be grouped into
   a glycolysis and the Krebs cycle
   b the Krebs cycle and photophosphorylation
   c the light reactions and the dark reactions
   d the light reactions and glycolysis

8. Chlorophyll is present
   a on the surface of chloroplasts
   b dispersed throughout the chloroplasts
   c in the stroma of chloroplasts
   d in the grana of chloroplasts

9. Hydrogen is transferred from the light reactions to the dark reactions by
   a NADP
   b DPN
   c ATP
   d DNA

10. The function of ATP in photosynthesis is the transfer of energy from the
   a dark reactions to the light reactions
   b light reactions to the dark reactions
   c chloroplasts to the mitochondria
   d mitochondria to the chloroplasts

1 ______
2 ______
3 ______
4 ______
5 ______
6 ______
7 ______
8 ______
9 ______
10 ______

BASIC FACTS

PHOTOSYNTHESIS is the process by which plants convert light into the chemical energy stored in foods. In photosynthesis carbon dioxide and water react in the presence of light and chlorophyll to form sugar and oxygen.

CARBON DIOXIDE and WATER are the raw materials; they are low-energy, inorganic compounds that are converted into high-energy, organic compounds (sugars). LIGHT is the energy source for the reaction. CHLOROPHYLL is required for its ability to absorb light. Violet, blue, and red light are the components of visible light that are most efficiently absorbed by chlorophyll. Chlorophyll is present in GRANA, small, disc-shaped bodies in the chloroplasts.

A number of conditions influence the RATE OF PHOTOSYNTHESIS. Within limits, the higher the concentration of CARBON DIOXIDE and the higher the INTENSITY OF LIGHT, the higher is the rate of photosynthesis, but extremely high concentrations of carbon dioxide and intensities of light will inhibit the process. WATER is ordinarily present in ample quantities, but if the water content of the plant becomes so low that the plant wilts, the stomata close, and the supply of carbon dioxide is shut off with a resultant decrease in photosynthetic rate. The optimum TEMPERATURE for photosynthesis varies with the species. For the majority of plants it lies between 10°C and 35°C.

Photosynthesis is divided into the LIGHT REACTIONS, in which water is split into hydrogen and oxygen, and the DARK REACTIONS, in which the hydrogen converts carbon dioxide to carbohydrates. The light reactions are rapid and require light directly; the dark reactions are slow and can take place in the absence of light. The light required in the light reactions is used in the formation of ATP from ADP

(Continued on page 20)

ADDITIONAL INFORMATION

In living cells energy is transferred by the compound ADENOSINE TRIPHOSPHATE (ATP). An ATP molecule contains three phosphate groups. It is synthesized in living things by the reaction of adenosine diphosphate (ADP) with phosphoric acid (H_3PO_4). The chemical bond between the two terminal phosphates is a high-energy bond. Energy is required to make the bond, and that energy is released when the bond is broken:

$$\text{ADP} + H_3PO_4 + \text{energy} \rightarrow \text{ATP} \quad (1)$$

$$\text{ATP} \rightarrow \text{ADP} + H_3PO_4 + \text{energy} \quad (2)$$

In photosynthesis ATP is produced in the light reactions and used in the dark reactions.

OXIDATION is usually defined as the removal of electrons from an atom, and REDUCTION as the addition of electrons to an atom. Oxidation may also be considered as the addition of oxygen to, or the removal of hydrogen from, an atom, and reduction as the removal of oxygen from, or the addition of hydrogen to, an atom. Thus in photosynthesis when $6CO_2$ is converted to $C_6H_{12}O_6$ the carbon atoms are reduced, for hydrogen has been added to them, and some oxygen has been removed. One of the molecules that commonly transfer hydrogen in living cells is NICOTINAMIDE ADENINE DINUCLEOTIDE PHOSPHATE (NADP), also called TRIPHOSPHOPYRIDINE NUCLEOTIDE (TPN). It is capable of taking up two hydrogen atoms to become $NADPH_2$ ($TPNH_2$). $NADPH_2$ reacts with the substance to be reduced, transferring the hydrogen atoms to it and thereby becoming NADP again. $NADPH_2$ is produced in the light reactions and is used in the dark reactions. Reductions require energy, and oxidations release energy.

Two LIGHT REACTIONS occur in green plants: cyclic photophosphorylation and noncyclic photophosphorylation. In CYCLIC PHOTOPHOSPHORYLATION light is absorbed by a chlorophyll molecule. Some of the light energy is absorbed by one of the electrons in the molecule. The electron becomes a high-energy electron and is transferred to a series of molecules called ELECTRON ACCEPTORS which return it to the chlorophyll molecule, thus restoring it to its original low-energy state. The identity of all the acceptors is not known with certainty, but some of them belong to the cytochrome system. As the electron passes along this series of molecules, its extra energy is lost and used in the synthesis of ATP. In NONCYCLIC PHOTOPHOSPHORYLATION both ATP and $NADPH_2$ are produced. A high-energy electron from an illuminated chlorophyll molecule is passed by a series of electron acceptors to a second chlorophyll molecule. In the transfer the electron loses energy that is used in the synthesis of ATP. Light striking the second chlorophyll molecule is absorbed by it, and another high-energy electron is released and carried by another electron acceptor to a molecule of

(Continued on page 20)

EXPLANATIONS

1. *The rate of photosynthesis increases when* the temperature is increased to a certain limit. The optimum temperature varies with the species. The rate of photosynthesis increases with an increase in carbon dioxide concentration and with an increase in light intensity, but when the carbon dioxide levels are extremely high or light is extremely intense, the rate is retarded. An increase in water content will increase the photosynthetic rate only if the plant has begun to wilt.

2. *In photosynthesis, light* is converted into chemical energy in food molecules. In a chemical reaction one form of energy may be converted into another form of energy. When wood burns chemical energy is converted into heat and light. Photosynthesis is a unique process in the biological world because it converts light into a form of energy that can be stored for months or years and can be used by night as well as by day.

3. *Before light can be used in photosynthesis it must be* absorbed by the green pigment chlorophyll.

4. *Light that is most effective in photosynthesis in green plants is* red, blue, and violet, the wavelengths that are absorbed most efficiently by chlorophyll.

5. *When a compound gains electrons or hydrogen atoms it is said to be* reduced. It is oxidized by the loss of electrons or hydrogen atoms.

6. *The role of water in photosynthesis is* providing hydrogen with which carbon dioxide is reduced to glucose. The oxygen that is a by-product of photosynthesis also comes from water. Chlorophyll is not soluble in water.

7. *The steps in photosynthesis may be grouped into* the light reactions, which require light, and the dark reactions, which do not require light. The dark reactions, however, require that the light reactions immediately precede them, so in practice both occur in light.

8. *Chlorophyll is present* in the grana of chloroplasts where the light reactions occur. The dark reactions occur in the stroma.

9. *Hydrogen is transferred from the light reactions to the dark reactions by* NADP, a common hydrogen carrier in biological systems.

10. *The function of ATP in photosynthesis is the transfer of energy from the* light reactions to the dark reactions. In the light reactions light energy is converted into the chemical energy of the phosphate bonds in ATP molecules. In the dark reactions this energy is used in the reduction of carbon dioxide to glucose.

Answers

d	1
b	2
c	3
c	4
c	5
b	6
c	7
d	8
a	9
b	10

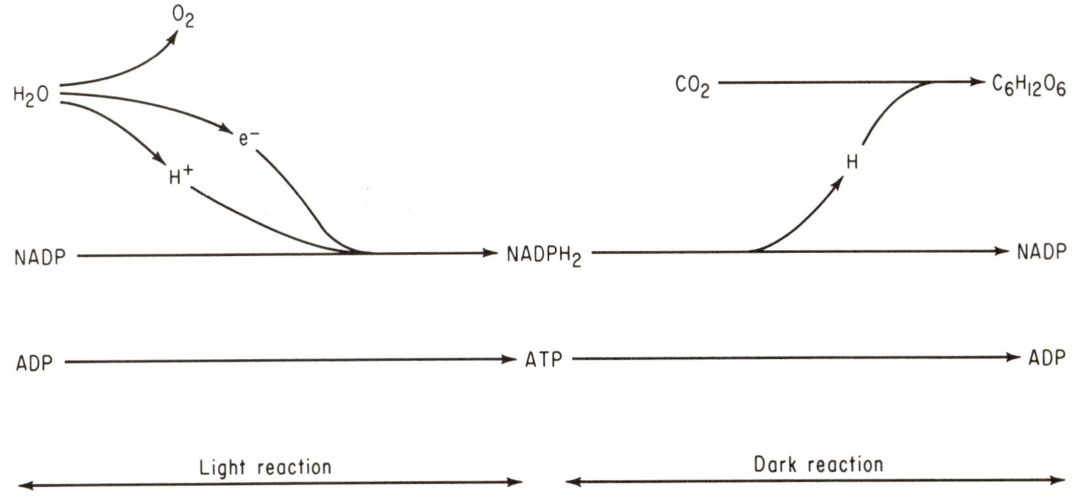

Summary of Photosynthesis

and $NADPH_2$ from NADP. The energy transferred to these compounds is then used in the dark reactions.

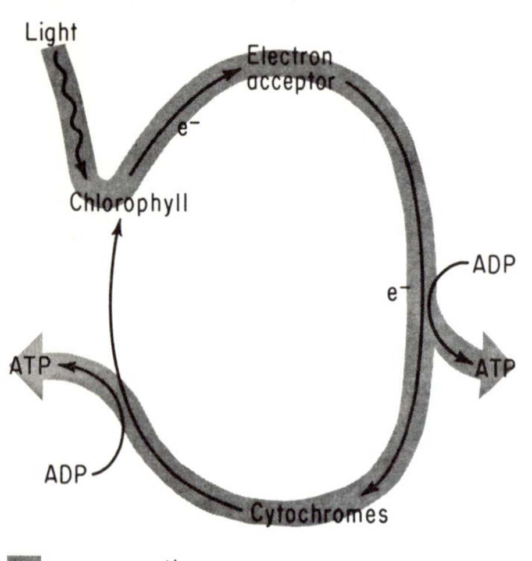

Cyclic Photophosphorylation

NADP. Water molecules ionize, forming hydrogen ions (H^+) and hydroxyl ions (OH^-). The OH^- releases one electron to the first chlorophyll molecule, replacing the one that was lost; the remainder of the ion forms O_2 and H_2O. The hydrogen ions combine with NADP and electrons from the second chlorophyll molecule, forming $NADPH_2$.

In the DARK REACTIONS each molecule of carbon dioxide reacts with a molecule of ribulose diphosphate, forming two 3-carbon molecules of phosphoglyceric acid which are then reduced to phosphoglyceraldehyde (a phosphorylated 3-carbon sugar) by reaction with ATP and $NADPH_2$. ATP is converted to ADP, $NADPH_2$ to NADP. Two phosphoglyceraldehyde molecules form a fructose diphosphate molecule which can be converted to glucose; other phosphoglyceraldehyde and fructose diphosphate molecules regenerate molecules of ribulose diphosphate.

A summary equation for photosynthesis is sometimes written:

$$6CO_2 + 6H_2O \rightarrow C_6H_{12}O_6 + 6O_2$$

Experiments using heavy oxygen as a tracer have indicated that water is the source of all the oxygen produced in photosynthesis. Since six water molecules do not contain enough oxygen atoms to account for all of the molecular oxygen produced, six additional water molecules are usually added to both sides of the equation:

$$6CO_2 + 12H_2O \rightarrow C_6H_{12}O_6 + 6O_2 + 6H_2O$$

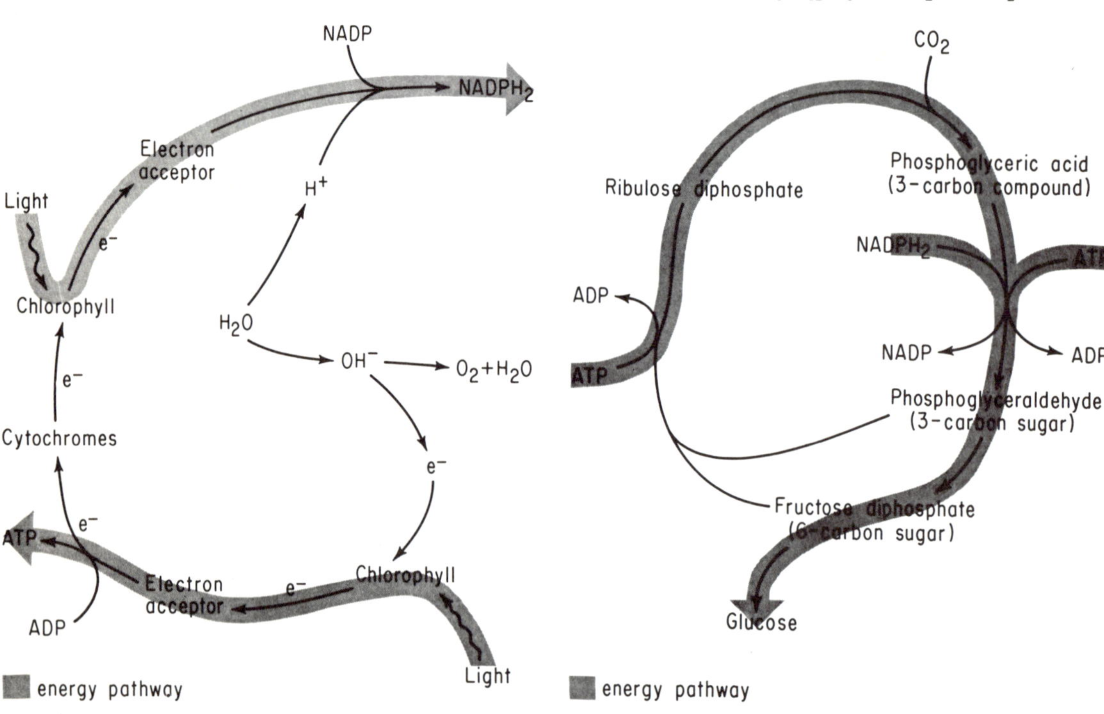

Noncyclic Photophosphorylation

Dark Reaction

6 DIGESTION AND RESPIRATION

SELF-TEST

1. Digestion is the conversion of
   - a simple organic compounds into complex organic compounds
   - b complex organic compounds into simple organic compounds
   - c simple organic compounds into inorganic compounds
   - d inorganic compounds into simple organic compounds

2. The energy in foods is made available to cells by the process of
   - a reduction
   - b digestion
   - c respiration
   - d condensation

3. Respiration occurs
   - a in all living cells but only in the light
   - b in all living cells in both light and dark
   - c only in nongreen cells in the light
   - d only in nongreen cells but in both light and dark

4. Aerobic respiration involves the
   - a oxidation of glucose by the removal of hydrogen from it
   - b reduction of glucose by the removal of oxygen from it
   - c reduction of glucose by the addition of oxygen to it
   - d oxidation of glucose to ethyl alcohol

5. Respiration results in the
   - a synthesis of ATP and the destruction of NAD
   - b destruction of ATP and synthesis of NAD
   - c synthesis of ATP
   - d synthesis of NAD

6. Most of the ATP synthesized in aerobic respiration is formed in
   - a glycolysis
   - b the Krebs cycle
   - c terminal oxidation

7. The conversion of glucose to pyruvic acid occurs in
   - a aerobic respiration, the light reaction, and alcoholic fermentation
   - b the dark reaction, anaerobic fermentation, and lactic acid fermentation
   - c light reaction, aerobic respiration, and lactic acid fermentation
   - d aerobic respiration, alcoholic fermentation, and lactic acid fermentation

8. The Krebs cycle occurs in
   - a ribosomes
   - b endoplasmic reticulum
   - c mitochondria
   - d Golgi apparatus

9. Free oxygen is required by
   - a all plants
   - b most plants
   - c few plants
   - d no plants

1 ________
2 ________
3 ________
4 ________
5 ________
6 ________
7 ________
8 ________
9 ________

BASIC FACTS

The chemical energy stored in foods is released in usable form by their OXIDATION within the cell. This process is called RESPIRATION. Large, complex organic molecules such as starch, fat, and proteins cannot be oxidized directly, nor can they diffuse through cell membranes. They must first be DIGESTED, or broken down into simpler, soluble compounds. With the aid of specific enzymes, starch is digested to maltose, maltose to glucose, fats to fatty acids and glycerol, and proteins to amino acids. Chemically, digestion is a HYDROLYSIS reaction, one in which a molecule is split into smaller molecules by the addition of water. Some energy is released, but it is in the form of heat and is not useful to the cell. Digestion in most plants is INTRACELLULAR.

AEROBIC RESPIRATION is the oxidation of glucose (or other simple foods), in the presence of free oxygen, to carbon dioxide and water. Burning glucose in a flame is also an oxidation. The two processes have the same overall equation:

$$C_6H_{12}O_6 + 6O_2 \rightarrow 6CO_2 + 6H_2O$$

However, burning is a rapid process that releases energy as light and heat; such high temperatures are attained that living cells would be destroyed. Respiration is a series of reactions that occur in a stepwise fashion at ordinary temperatures. Energy is released slowly, and a considerable amount is transferred to ATP molecules. This energy "trapped" by ATP can be used by the cells for their many energy-requiring activities. Each of the separate respiration reactions is controlled by a specific enzyme. Enzymatic action lowers the heat requirement for the initiation of each step.

Aerobic respiration is divided into GLYCOLYSIS, in which glucose is converted to pyruvic acid by the removal of hydrogen; the KREBS CYCLE (CITRIC

(Continued on page 24)

ADDITIONAL INFORMATION

AEROBIC RESPIRATION is a series of chemical reactions that are usually divided into three groups: glycolysis, Krebs cycle, and terminal oxidation. In these reactions hydrogen atoms are removed from glucose which is thereby converted to carbon dioxide. In aerobic respiration the hydrogen carrier is NICOTINAMIDE ADENINE DINUCLEOTIDE (NAD), which is also called DIPHOSPHOPYRIDINE NUCLEOTIDE (DPN). NAD transfers the hydrogen to free oxygen, thus forming water.

In GLYCOLYSIS a series of chemical reactions splits one molecule of glucose into two molecules of pyruvic acid. In the process two NAD molecules take up four hydrogen atoms removed in the oxidation of glucose and become $NADH_2$, and two ATP molecules are formed. Glycolysis occurs in the clear portion of the cytoplasm.

In the KREBS CYCLE the pyruvic acid formed in glycolysis is broken down into carbon dioxide. Before entering the cycle, two more hydrogen atoms and a carbon dioxide molecule are removed from each pyruvic acid molecule, thus converting it into a two-carbon fragment. This fragment combines with COENZYME A (CoA), forming acetyl CoA, which enters the Krebs cycle. Each acetyl CoA molecule reacts with a four-carbon compound, oxaloacetic acid. In the reaction the two-carbon fragment is added to oxaloacetic acid, forming a six-carbon compound, CITRIC ACID, and CoA is released. Another series of chemical reactions converts citric acid back to oxaloacetic acid. In this latter series, for each citric acid molecule, two molecules of carbon dioxide are released, and eight atoms of hydrogen are taken up by four NAD molecules. A balanced equation for glycolysis and the Krebs cycle is:

$$C_6H_{12}O_6 + 12NAD + 6H_2O \rightarrow 6CO_2 + 12NADH_2 \quad (1)$$

In TERMINAL OXIDATION, $NADH_2$ releases hydrogen atoms which ionize, their electrons being transferred to cytochrome molecules. As the electrons pass from one cytochrome molecule to another, they lose some of their energy; this energy is used in the synthesis of ATP. Finally, the hydrogen ions and electrons combine with oxygen, forming water. Ignoring the formation of ATP, we may summarize the terminal oxidation:

$$12NADH_2 + 6O_2 \rightarrow 12NAD + 12H_2O \quad (2)$$

If we add equations (1) and (2) and cancel out the molecules that appear on both sides of the equation we get the overall equation for aerobic respiration:

$$C_6H_{12}O_6 + 6O_2 \rightarrow 6CO_2 + 6H_2O$$

The chemical energy of the glucose molecule is transferred to the ATP molecules that are formed in terminal oxidation. This synthesis of ATP is called OXIDATIVE PHOSPHORYLATION. A total of 38 ATP molecules are formed for each glucose molecule completely oxidized. When ATP is converted

(Continued on page 24)

EXPLANATIONS

1. *Digestion is the conversion of* complex organic compounds into simple organic compounds. Large food molecules must be digested before they can be oxidized or transported, for they are generally too large to pass through plasma membranes.

2. *The energy in foods is made available to cells by the process of* respiration. The chemical energy of the food molecules is transferred to ATP which in turn transfers it to energy-requiring reactions.

3. *Respiration occurs* in all living cells in both light and dark. All living things constantly utilize energy which they obtain from their food by respiration. When a cell stops respiring, it dies.

4. *Aerobic respiration involves the* oxidation of glucose (and a few other foods) by the removal of hydrogen from it. A few, but not all, of the individual chemical reactions are the reverse of those of photosynthesis.

5. *Respiration results in the* synthesis of ATP. NAD serves as a hydrogen carrier and is unchanged when the individual chemical reactions of respiration are completed.

6. *Most of the ATP synthesized in aerobic respiration is formed in* terminal oxidation. For every molecule of glucose oxidized, 36 molecules of ATP are formed in terminal oxidation after completion of the Krebs cycle, and only 2 are formed in glycolysis.

7. *The conversion of glucose to pyruvic acid occurs in* aerobic respiration, alcoholic fermentation, and lactic acid fermentation. Pyruvic acid is an important compound in both aerobic and anaerobic respiration. Under aerobic conditions it may be converted to carbon dioxide and water by the reactions of the Krebs cycle. Under anaerobic conditions it may be converted to any one of several organic compounds by fermentations.

8. *The Krebs cycle occurs in* mitochondria, small granular or rod-shaped bodies in the cytoplasm.

9. *Free oxygen is required by* most plants. Many plants can survive without free oxygen for some time, but most of them will not live long without it.

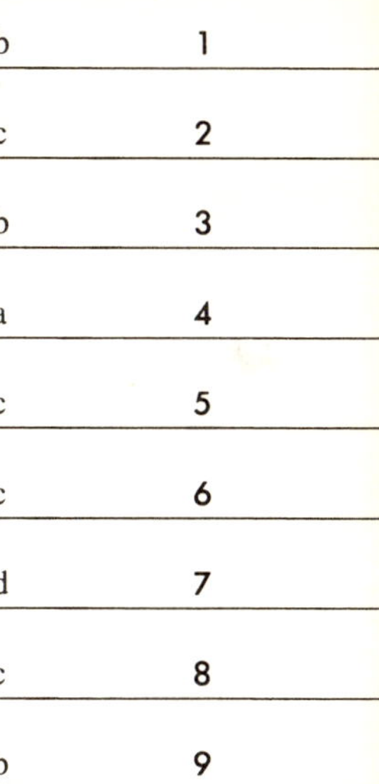

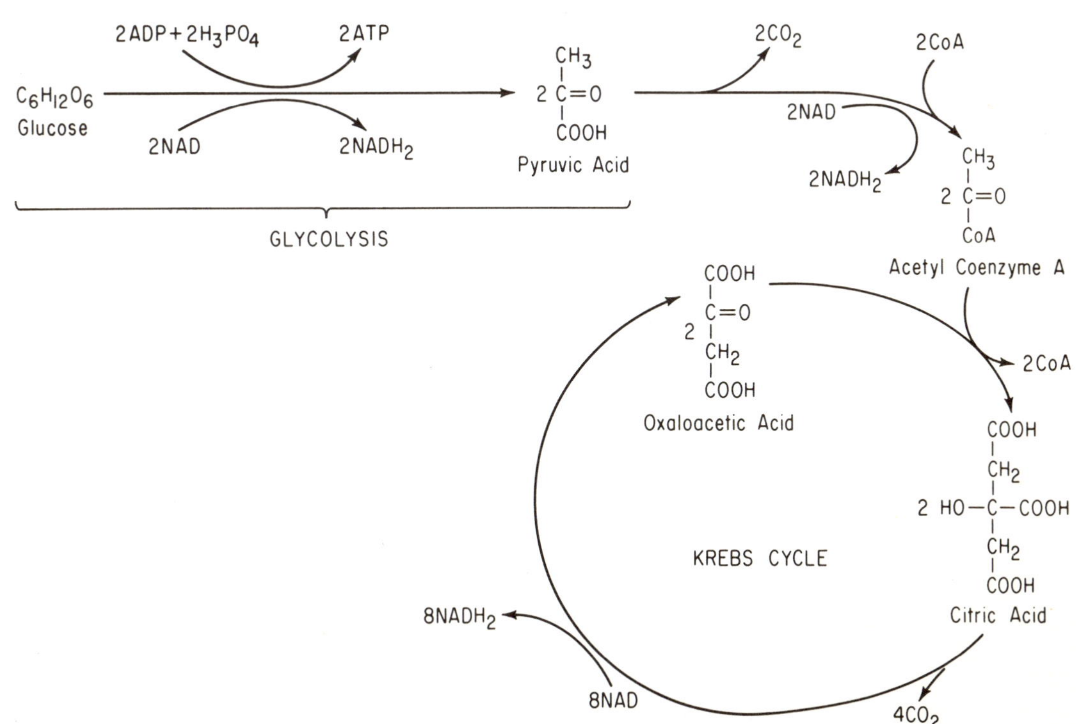

Glycolysis and the Krebs Cycle

ACID CYCLE), in which pyruvic acid is converted to carbon dioxide by the removal of more hydrogen; and TERMINAL OXIDATION, in which the hydrogen combines with molecular oxygen, thus forming water.

FERMENTATIONS do not require molecular oxygen and they release much less energy than does aerobic respiration. In fermentation, glucose is converted into another simple organic compound. Ethyl alcohol and carbon dioxide are the products of ALCOHOLIC FERMENTATION, an anaerobic process carried on by yeasts.

The RATE of respiration is affected by a number of environmental factors. The amount of GLUCOSE available is usually adequate in plants carrying out normal photosynthesis, and the CONCENTRATION OF OXYGEN in the air is usually more than ample to support normal aerobic respiration. Within limits an increase in TEMPERATURE causes an increase in the respiration rate, but very high temperatures coagulate enzymes and cause respiration to slow down or to stop. The amount of WATER in tissues also affects the rate of respiration. Very dry tissues such as dormant seeds or spores have very low rates, but moist, active tissues have high rates.

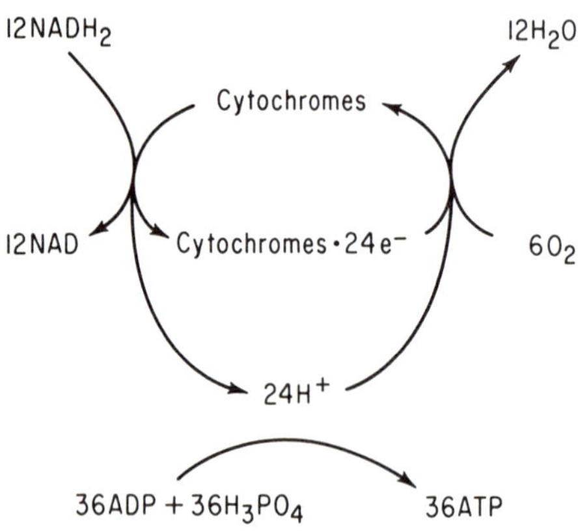

Terminal Oxidation

to ADP its energy is used in all energy-requiring reactions of living cells (except for the light reactions of photosynthesis, in which light is the energy used).

The Krebs cycle and terminal oxidation occur in MITOCHONDRIA. Each mitochondrion is bounded by a double membrane, the inner membrane bearing folds called CRISTAE. It is on these cristae that the enzymes of the Krebs cycle and terminal oxidation are located.

FERMENTATION takes place when molecular oxygen is not available. The first steps are identical with those of glycolysis. Pyruvic acid is then converted to ethyl alcohol, lactic acid, or another organic compound. In ALCOHOLIC FERMENTATION pyruvic acid is converted into acetaldehyde which is then reduced to ethyl alcohol by the $NADH_2$ produced in glycolysis:

$$2\ \underset{\text{COOH}}{\overset{\text{CH}_3}{\text{C}}}{=}\text{O} \xrightarrow{\ 2CO_2\ } 2\ \text{H}{-}\overset{\text{CH}_3}{\text{C}}{=}\text{O} \xrightarrow{\ 2NADH_2 \ \to\ 2NAD\ } 2\ \text{H}_2\overset{\text{CH}_3}{\text{C}}{-}\text{OH}$$

Pyruvic acid — Acetaldehyde — Ethyl alcohol

Thus one glucose molecule is converted into two carbon dioxide molecules and two ethyl alcohol molecules. The amount of energy released by alcoholic fermentation is much less than that produced by aerobic respiration. Only two molecules of ATP are produced in contrast to 38 in aerobic respiration.

Comparison of Photosynthesis and Aerobic Respiration

PHOTOSYNTHESIS	RESPIRATION
Carbon dioxide and water are raw materials	Glucose and oxygen are raw materials
Glucose and oxygen are produced	Carbon dioxide and water are produced
Reduction reaction	Oxidation reaction
ATP produced in light reaction is used in dark reaction	ATP produced in terminal oxidation is used in various chemical reactions in the cell
Occurs only in chlorophyll-containing cells	Occurs in all living cells
Occurs only in light	Occurs in light and darkness
Occurs in chloroplasts	Occurs in mitochondria

7 ROOTS

SELF-TEST

1. The primary functions of the root are
   a conduction and storage
   b storage and anchorage
   c anchorage and absorption
   d absorption and conduction

2. A branch root is a(n)
   a secondary root
   b adventitious root
   c tap root
   d aerial root

3. Beginning at the root tip the order of the regions of the root are:
   a root cap, apical meristem, region of elongation, and region of maturation
   b root cap, apical meristem, region of maturation, and region of elongation
   c apical meristem, root cap, region of elongation, and region of maturation.
   d apical meristem, root cap, region of maturation, and region of elongation

4. Root hairs occur in the
   a root cap
   b region of cell elongation
   c apical meristem
   d region of maturation

5. The primary meristem of the root is the
   a apical meristem
   b vascular cambium
   c cork cambium
   d pericycle

6. The secondary meristems of the root are
   a cork cambium and apical meristem
   b apical meristem and pericycle
   c pericycle and vascular cambium
   d vascular cambium and cork cambium

7. Vascular cambium produces
   a primary and secondary xylem
   b primary and secondary phloem
   c primary xylem and primary phloem
   d secondary xylem and secondary phloem

8. The pericycle gives rise to
   a cortex and pith
   b epidermis and vascular cambium
   c branch roots and cork cambium
   d xylem and phloem

9. Monocot roots generally differ from dicot roots in having
   a pith but no cambium
   b endodermis but no pericycle
   c both vascular cambium and cork cambium
   d both primary and secondary growth

10. A mycorhiza is a
    a long, thin root
    b rootlike, underground stem
    c parasitic root
    d combination of a root and a fungus

11. Prop roots are
    a primary roots
    b adventitious roots
    c branch roots
    d tap roots

1 ______
2 ______
3 ______
4 ______
5 ______
6 ______
7 ______
8 ______
9 ______
10 ______
11 ______

BASIC FACTS

The root system of a plant anchors it in the soil and absorbs water and minerals. A TAP ROOT SYSTEM has one large primary root (the first root produced by an embryo) that is thicker than its branch, or secondary, roots. In a FIBROUS ROOT SYSTEM most of the roots are of about the same size. ADVENTITIOUS ROOTS are those that develop from organs other than roots.

Beginning at the root tip the regions of a young root are root cap, apical meristem, region of elongation, and region of maturation. The ROOT CAP protects the delicate apical meristem from damage as the root grows through the soil. As root cap cells are injured and die, they are replaced by cells from the apical meristem. The APICAL MERISTEM is a mass of embryonic cells which by their divisions provide new cells to the root. Some new cells remain in the apical meristem; others are contributed to the root cap and to the region of elongation. The REGION OF ELONGATION lies above the apical meristem. Elongation of cells in this region causes the apical meristem and the root cap to be pushed downward through the soil. The apical meristem and the region of elongation account for growth in length of the root. In the REGION OF MATURATION (ROOT-HAIR ZONE) cells begin to differentiate into the primary tissues of the root.

The PRIMARY TISSUES of the root are epidermis, cortex, endodermis, pericycle, primary phloem, primary xylem, and pith (if present). The EPIDERMIS is a single layer of cells that completely surrounds the young root. It provides protection and absorbs most of the water and minerals that enter the plant. ROOT HAIRS are extensions of epidermal cells low in the region of maturation. They account for most of the absorbing surface of the root. The CORTEX consists of thin-walled cells. Its primary functions are food storage and transport

(Continued on page 28)

ADDITIONAL INFORMATION

BRANCH ROOTS arise from the pericycle in the region of maturation. Small portions of the pericycle that abut on the primary xylem may become meristematic and undergo cell divisions. The new cells organize a new apical meristem which acts like that of the parent root and produces its own root cap, region of elongation, and region of maturation. As the branch root grows it forces its way through the endodermis, cortex, and epidermis and emerges from the parent root. The vascular tissues of the branch root are connected directly with the vascular tissues of the parent root. As branch roots grow older they produce their own branch roots.

When a VASCULAR CAMBIUM cell divides, the new cell wall between the two daughter cells is parallel to the surface of the root. If the inner daughter cell matures into a secondary xylem cell, the outer daughter cell becomes a vascular cambium cell. If the outer daughter cell matures into a secondary phloem cell, the inner daughter cell becomes a vascular cambium cell. Repeated divisions produce rings of secondary xylem (between primary xylem and vascular cambium) and secondary phloem (between vascular cambium and primary phloem). They also maintain the vascular cambium indefinitely. Since more divisions produce new xylem than new phloem, there is usually more secondary xylem than secondary phloem. Vascular cambium continues to function throughout the life of the root. The regular production of secondary xylem gradually pushes the vascular cambium farther and farther away from the center of the root.

Divisions of CORK CAMBIUM cells are similar to those of vascular cambium cells. The outer daughter cell usually matures into a cork cell, and the inner one becomes a cork cambium cell. Rarely the outer cell becomes a cork cambium cell and the inner a phelloderm cell. Repeated divisions produce rings of cork and phelloderm. They also maintain the cork cambium. Tissues external to the pericycle die, and cork then assumes the protective functions of the epidermis. The oldest (outermost) cork cells usually die and decay or slough off. They are replaced by the continued activity of the cork cambium. Cork often forms on the surfaces of wounded tissues.

Roots frequently enter into SYMBIOTIC RELATIONSHIPS with bacteria and fungi. *Rhizobium* is a nitrogen-fixing species of BACTERIA that converts free nitrogen (N_2), which green plants cannot use, into a usable form. The bacteria infect the roots of leguminous plants (bean, peas, clover, alfalfa) and cause the formation of swellings, or nodules, on the roots. The bacteria live in the nodules and obtain their food from the host plant, which in turn absorbs some of the fixed nitrogen. Some FUNGI grow in symbiotic combinations called MYCORHIZAE with the roots of higher plants. The fungus is thought to obtain food from the host plant. In some cases

(Continued on page 28)

EXPLANATIONS

1. *The primary functions of the root are* anchorage of the plant and absorption of water and minerals. Conduction and storage are functions carried on by other organs as well as by roots.

2. *A branch root is a* secondary root, or lateral root, which arises from another root. Adventitious roots arise from organs other than roots. Aerial roots may or may not be branch roots; many of them are adventitious roots.

3. *Beginning at the root tip the order of the regions of the root are* root cap, apical meristem, region of elongation, and region of maturation. The apical meristem is the ultimate source of all the cells of the root.

4. *Root hairs occur in the* lower part of the region of maturation. If they were formed in the apical meristem, the root cap, or the region of elongation, they would be damaged by being dragged against soil particles as these parts are pushed downward.

5. *The primary meristem of the root is the* apical meristem located near the tip of the root. It provides for increase in length of the root. The cells it produces are typically arranged in longitudinal rows.

6. *The secondary meristems of the root are* the vascular cambium and the cork cambium. They provide for increase in width of the root. The cells they produce are typically arranged in radial rows.

7. *Vascular cambium produces* secondary xylem and secondary phloem. Primary xylem and primary phloem are produced by the apical meristem.

8. *The pericycle gives rise to* branch roots in young roots and to cork cambium in somewhat older roots. It also contributes to the formation of a small part of the vascular cambium, but most of the vascular cambium arises from undifferentiated tissue that lies between xylem and phloem.

9. *Monocot roots generally differ from dicot roots in having* pith but no cambium. The majority of monocots are annual plants and have no secondary growth and no cambium. Pith occupies the center of most monocot roots, but it is absent from dicot roots.

10. *A mycorhiza is a* symbiotic combination of a root and a fungus. Both members appear to benefit from the relationship.

11. *Prop roots are* adventitious roots which arise from stems and provide part of the support of the plant.

Answers

c	1
a	2
a	3
d	4
a	5
d	6
d	7
c	8
a	9
d	10
b	11

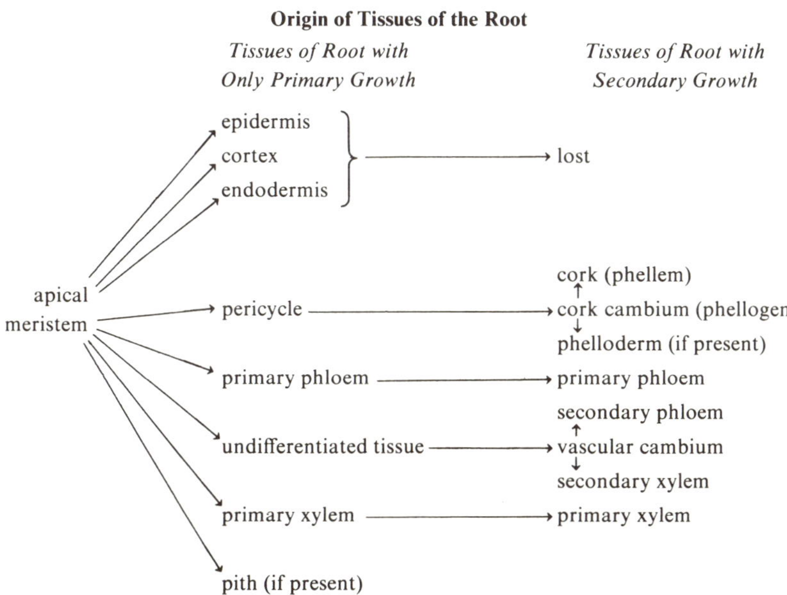

of water from root hairs to the central tissues of the root. The ENDODERMIS is a single layer of cells that surrounds the central tissues and that controls the entrance of water and dissolved materials into the xylem. Narrow bands of suberin (CASPARIAN STRIPS) line the upper, lower, and radial walls of each endodermal cell; because suberin is impermeable to water, dissolved substances cannot diffuse between the endodermal cells but must pass through the protoplasts before entering the vascular tissue. The PERICYCLE, usually a single layer of cells, lies between the endodermis and the vascular tissue. It gives rise to branch roots and cork cambium. The PRIMARY PHLOEM of the root exists as separate strands arranged in a circle within the pericycle. The PRIMARY XYLEM of dicot roots occupies the center of the root. In cross section it appears star-shaped with the arms of the star occupying the spaces between phloem strands. In monocot roots the xylem occurs as separate strands which alternate in position with the phloem strands, and PITH occupies the center of the root. The STELE, or VASCULAR CYLINDER, includes the primary xylem, primary phloem, and pith (if present); some botanists also consider the pericycle as part of the stele.

SECONDARY TISSUES of the root are SECONDARY XYLEM and SECONDARY PHLOEM, which are produced by the vascular cambium, and CORK (phellem) and PHELLODERM (a parenchyma-like tissue), which are produced by the cork cambium (phellogen). VASCULAR CAMBIUM is a single layer of cells that arise from a narrow band of cells that remain as undifferentiated tissue between the primary xylem and primary phloem in young perennial roots. CORK CAMBIUM arises from the pericycle.

the fungi act like root hairs and increase the amount of water absorbed by the host plant. Some may digest organic compounds in the soil, thus making certain nitrogen-rich compounds more readily available to the host plant. Some fungi form a mat covering the surface of the root and penetrating the intercellular spaces among the outer cells; others grow within the root cells.

Roots may be modified to perform functions other than absorption and anchorage. STORAGE ROOTS such as those of carrots, beets, and dahlias contain a great deal of food, usually carbohydrates. PROP ROOTS are adventitious roots which are produced by the stem some distance above the ground; they grow outward and downward until they penetrate the soil, thus furnishing additional support to the plant. Prop roots occur in corn and certain tropical trees. Adventitious AERIAL ROOTS of climbing vines, such as ivy, also provide support. The aerial roots of tropical epiphytes may have a spongy layer of epidermal cells, the velamen, which absorbs rain or dew. Roots of certain aerial orchids are green and carry on PHOTOSYNTHESIS. HAUSTORIA are the roots of parasitic plants. They penetrate the host plant and draw water and food from it. The roots of a few plants are involved in ASEXUAL REPRODUCTION. Adventitious buds appear on the roots and give rise to new shoots. Sweet potatoes commonly reproduce this way as do a number of weeds such as dandelions.

8 DIFFUSION AND OSMOSIS

SELF-TEST

1. Diffusion will not take place unless
   a the molecules of a substance are in rapid motion
   b the diffusion pressure of a substance varies within the available space
   c ATP is present to provide the necessary energy
   d a differentially permeable membrane separates the two substances

2. In any mixture of two compounds
   a only one compound can diffuse at a time
   b both compounds diffuse in the same direction
   c both compounds diffuse in opposite directions
   d either, both, or none of the compounds may be diffusing, and in any direction

3. If a plant cell is immersed in water, water will continue to enter the cell
   a until the concentration of salts is the same inside the cell as outside
   b until the cell bursts
   c until the diffusion pressure is the same inside the cell as outside
   d until the concentration of water is the same inside the cell as outside

4. A substance that would be expected to enter a plant by osmosis is
   a water
   b carbon dioxide
   c magnesium

5. Guttation is
   a the evaporation of water vapor from the stomata
   b the absorption of water through the roots
   c the storage of water by fleshy stems
   d the loss of liquid water from leaves

6. Plasmolysis occurs if a cell is placed in a solution that has
   a a higher diffusion pressure than that of the cell
   b a lower diffusion pressure than that of the cell
   c the same diffusion pressure as the cell

7. When a plant cell is plasmolyzed, the space between the plasma membrane and the cell wall is occupied by
   a the plasmolyzing solution
   b water
   c air
   d the original vacuolar material

8. Transpiration and root pressure cause water to rise in plants by
   a pushing it upward
   b pulling it upward
   c pulling and pushing it, respectively
   d pushing and pulling it, respectively

9. The rise of water to the tops of tall trees is accounted for by
   a transpirational pull
   b osmosis
   c atmospheric pressure
   d capillarity

10. The opening and closing of stomata are influenced by
    a root pressure
    b osmosis
    c guttation
    d imbibition

11. The rate of transpiration is affected by
    a temperature
    b relative humidity
    c wind
    d all of the above

1 ________
2 ________
3 ________
4 ________
5 ________
6 ________
7 ________
8 ________
9 ________
10 ________
11 ________

BASIC FACTS

Much of the water entering a plant enters through the root by OSMOSIS. This ordinarily occurs when the concentration of water is greater outside the root than it is in the cells. Water enters primarily through the root hairs of the epidermis, for these account for most of the absorbing surface of the root. It moves across the cortex, the endodermis, and the pericycle, and then enters the xylem. Although water can move upward through any tissue, the xylem is the main channel of water transport. In the aerial portions of the plant it moves from the xylem to those tissues that require it, but most of the water evaporates from the mesophyll cells and diffuses out of the leaves through the stomata. This loss of water is called TRANSPIRATION. It performs no useful function in the plant, and if it exceeds absorption of water for a prolonged time, it may be harmful because it may cause wilting and even death.

If absorption exceeds transpiration, a pressure, known as ROOT PRESSURE, develops in the roots and forces water upward some distance in the plant. In some species of plants root pressure is responsible for GUTTATION, the exuding of droplets of water from the leaves. These plants have HYDATHODES, which are openings at the ends of the vascular bundles of the leaves; these openings resemble stomata, but do not have guard cells and are permanently open. When root pressure is high, BLEEDING, the exuding of water from wounded tissue, may occur.

The evaporation of water from the leaves creates a tension in the water column in the xylem and "pulls" water upward from the root. Most of the water rising in plants is pulled upward by transpiration rather than being pushed upward by root pressure and osmosis.

(Continued on page 32)

ADDITIONAL INFORMATION

Individual molecules of all substances are always in motion, moving in all directions. This molecular activity is called DIFFUSION PRESSURE. The diffusion pressure of a substance can be increased by increasing the concentration of the substance, by raising its temperature, or by applying pressure to it. DIFFUSION is the net movement of molecules of a particular substance from a region of high diffusion pressure of that substance to a region of low diffusion pressure of that substance. Diffusion continues until there is no difference in diffusion pressure in the space available. Some substances enter and leave plants by simple diffusion.

During the day when the rate of photosynthesis is higher than that of respiration and carbon dioxide is being used by the leaf faster than it is being produced, the carbon dioxide concentration in the leaf is lower than it is in the surrounding atmosphere. Carbon dioxide then diffuses into the leaf through the stomata. The oxygen concentration is higher in the leaf than in the surrounding atmosphere, and oxygen diffuses out of the leaf. At night when no photosynthesis occurs, the relationships are reversed, and carbon dioxide diffuses out of the leaf and oxygen diffuses in. Diffusion requires no expenditure of energy by the plant.

OSMOSIS is a special case of diffusion; it is diffusion of a solvent through a DIFFERENTIALLY PERMEABLE MEMBRANE. Such a membrane permits some substances to pass through it, but not others. The cellular membranes are differentially permeable, but for ease of discussion we may consider the entire cytoplasm as a single such membrane that separates the vacuole from the environment. The vacuoles contain sugars, salts, and other dissolved materials to which the cytoplasm is relatively impermeable. The cytoplasm does, however, permit the relatively free passage of water. If a root is placed in distilled water, the concentration of water (and therefore its diffusion pressure) is greater outside the root than it is in the vacuoles of the root hairs where the water is diluted by the dissolved substances. Water then moves across the cytoplasm and enters the vacuole.

Although soil water contains some dissolved materials, the diffusion pressure of water around root hairs is ordinarily higher than it is in the root hairs. As water enters the cell, TURGOR PRESSURE builds up against the cell wall. This turgor pressure increases the diffusion pressure of the water within the cell. Osmosis continues until the diffusion pressure within the cell equals the diffusion pressure outside the cell. If soil water has a lower diffusion pressure than does the vacuole, water moves out of the cells by osmosis, and the cytoplasm shrinks and draws away from the cell walls. Such cells are said to be PLASMOLYZED. The cells become limp, or flaccid, and the plant wilts. Such solutions may be obtained by adding an excess of dissolved materials to the external solution. If the solution around the cells has the

(Continued on page 32)

EXPLANATIONS

1. *Diffusion will not take place unless* the diffusion pressure of a substance varies within the available space. When the diffusion pressure becomes equal throughout all the available space, diffusion stops. Changes in concentration of a substance, in temperature, or in external pressure will change diffusion pressure. Diffusion does not depend on the energy released by organisms; it may occur independently of living things.

2. *In any mixture of two compounds* either, both, or none of the compounds may be diffusing, and in any direction. The diffusion of one substance is independent of the diffusion of another. During the day oxygen and water vapor commonly diffuse out of leaves while carbon dioxide diffuses in.

3. *If a plant cell is immersed in water, water will continue to enter the cell* until the diffusion pressure is the same inside the cell as outside. As water enters the cell, turgor pressure develops within the cell and tends to force water out. When the turgor pressure equals the diffusion pressure of the external water, osmosis ceases. The concentration of water in the cell will still be lower than that outside.

4. *A substance that would be expected to enter a plant by osmosis is* water. Gases usually enter by simple diffusion, and minerals enter either by diffusion or by the expenditure of energy by the cell.

5. *Guttation is* the loss of liquid water from leaves. It occurs when the rate of transpiration is low and the rate of water absorption by the roots is high.

6. *Plasmolysis occurs if a cell is placed in a solution that has* a lower diffusion pressure than that of the cell. Water then leaves the cell, causing the cytoplasm to shrink away from the cell wall.

7. *When a plant cell is plasmolyzed, the space between the plasma membrane and the cell wall is occupied by* the plasmolyzing solution. The cell wall is permeable to both the water and dissolved substances in the solution which then occupies the space made available by the shrinking of the cytoplasm.

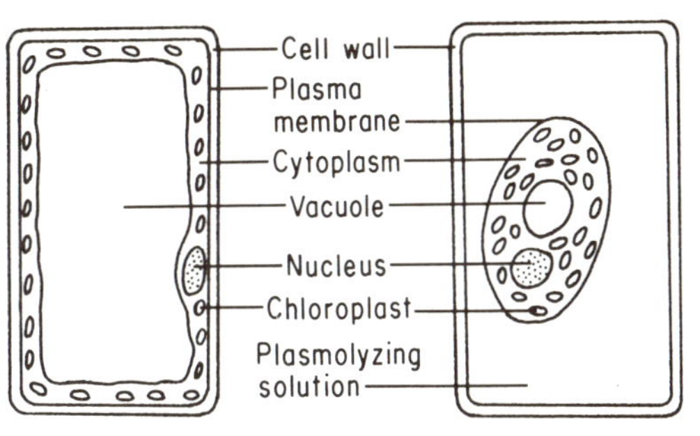

Turgid Cell Plasmolyzed Cell

8. *Transpiration and root pressure cause water to rise in plants by* pulling and pushing it, respectively. Loss of water by transpiration creates a tension from above, whereas root pressure is a pressure exerted from below.

9. *The rise of water to the tops of tall trees is accounted for by* transpirational pull. Under optimum conditions, osmosis, atmospheric pressure, and capillarity can account for the rise of water to the height of several feet, but not to the tops of trees that are several hundred feet high.

10. *The opening and closing of stomata are influenced by* the water content of the guard cells which is controlled by the entrance and exit of water into and out of these cells by osmosis.

11. *The rate of transpiration is affected by* all of the above. These factors change the diffusion pressure of water and water vapor and therefore affect the rate at which water evaporates and diffuses from the leaf.

Answers

Answer	No.
b	1
d	2
c	3
a	4
d	5
b	6
a	7
c	8
a	9
b	10
d	11

The RATE OF TRANSPIRATION is increased by an increase in temperature, by a decrease in relative humidity of the air, and by wind. But if any of these factors causes the loss of so much water from the guard cells that they wilt and close the stomata, the rate of transpiration decreases. The rate of transpiration is usually greatest in the afternoon when the temperature is high and the stomata are open, and lowest at night when the temperature is low and the stomata are closed.

The absorption of water that is due to osmosis is called ACTIVE ABSORPTION; that which is due to transpirational pull is called PASSIVE ABSORPTION.

The actual pressure exerted by the solutions within cells is TURGOR PRESSURE. It is responsible for the turgidity of cells and provides support for many organs that are not well supplied with xylem or other strengthening tissues. OSMOTIC PRESSURE is the potential pressure of a solution. It represents the turgor pressure that could be exerted by a solution if it were separated from pure water by a differentially permeable membrane. The higher the concentration of dissolved material in a solution the higher is its osmotic pressure.

IMBIBITION is the absorption of a liquid or a gas into the spaces between the molecules of a colloidal substance resulting in the swelling of that substance. Cellulose, starch, and proteins often imbibe water.

same diffusion pressure as that in the vacuoles, then the number of water molecules entering the cell is equal to the number leaving, and there is no diffusion and no osmosis.

CHANGES IN THE SIZE OF THE STOMATAL OPENINGS are caused by changes in the shape of the guard cells, which in turn are caused by changes in turgor. Guard cells are the only leaf epidermal cells that are photosynthetic. During the day they produce sugar, which lowers the diffusion pressure of water in the guard cells relative to that of their neighbors. Water then moves from the neighboring cells into the guard cells, thus increasing their turgidity. The portion of their walls that is adjacent to the stoma is relatively rigid and cannot stretch. As the guard cells become more and more turgid they are forced to assume a sausage shape that maintains an opening between them. At night when photosynthesis ceases, the sugar content is lowered and the diffusion pressure of water in the guard cells is raised relative to that of neighboring cells. Water moves from the guard cells into neighboring cells. As a result the guard cells become limp and collapse against each other, closing the stomata.

As a result of the pressure of water entering the root by osmosis, a certain amount of water may be forced upward in the xylem. This pressure is called ROOT PRESSURE. It is responsible for the rise of water in some plants in damp weather, but it is nonexistent when the soil is relatively dry, and the pressure is never sufficient to cause the rise of water to the tops of tall trees. According to the COHESION THEORY, water can rise to considerable heights as a result of transpirational "pull." The air spaces in leaves are saturated with water vapor; thus the diffusion pressure of water vapor in the leaf is ordinarily higher than that of the surrounding air. When stomata are open, water vapor diffuses out of the leaf (TRANSPIRATION). As water is lost from the intercellular spaces, it is replaced by water that evaporates from the cell walls of mesophyll cells, which in turn imbibe water from the cytoplasm of the mesophyll cells. This water is replaced by water that diffuses by osmosis from the xylem. Water has the property of cohesiveness, that is, its molecules tend to cohere. The degree of cohesion is especially high in narrow columns such as tracheids, with the result that the water is "pulled" up in the xylem, replacing the water that moves into the mesophyll.

9 MINERAL NUTRITION AND SOILS

SELF-TEST

1. Some minerals known to be required in large amounts for plant growth are
   a calcium, magnesium, manganese, copper
   b potassium, phosphorus, chlorine, boron
   c magnesium, sulfur, iron, zinc
   d phosphorus, potassium, sulfur, calcium

2. Magnesium is required by plants because it is part of
   a chlorophyll molecules
   b nucleic acids
   c phospholipids
   d proteins

3. Soil particles arranged in order of increasing size are
   a sand, silt, and clay
   b clay, silt, and sand
   c silt, sand, and clay
   d clay, sand, and silt

4. Clay soils are characterized by
   a small pore spaces and low field capacity
   b small pore spaces and high field capacity
   c large pore spaces and high field capacity
   d large pore spaces and low field capacity

5. A fertile soil is likely to have a pH of
   a 3
   b 8
   c 6 to 7
   d 11 to 12

6. Water that remains in the soil only temporarily after a rain
   a is gravitational water
   b is capillary water
   c represents field capacity
   d represents storage capacity

7. Soil water that is available to plants is
   a hygroscopic water and gravitational water
   b gravitational water and capillary water
   c capillary water and hygroscopic water

8. If the concentration of an element is higher in a plant than it is in the surrounding soil we may assume that
   a it is an essential element
   b it will soon become toxic to the plant
   c the plant rapidly oxidizes it
   d the plant expends energy when the element enters

9. A solution with a pH of 2 has a concentration of hydrogen ions that is
   a one-half that of a solution with a pH of 4
   b twice that of a solution with a pH of 4
   c ten times that of a solution with a pH of 4
   d one hundred times that of a solution with a pH of 4

10. Components of soils that are required by plants include
    a dissolved minerals, solid particles, and water
    b solid particles, water, and air
    c water, air, and dissolved minerals
    d air, dissolved minerals, and solid particles

1 ______

2 ______

3 ______

4 ______

5 ______

6 ______

7 ______

8 ______

9 ______

10 ______

BASIC FACTS

About fifteen elements are known to be required by plants. Ten of these (MACRONUTRIENTS) are needed in relatively large amounts: carbon, hydrogen, oxygen, nitrogen, phosphorus, potassium, sulfur, calcium, magnesium, and iron. Carbon, hydrogen, and oxygen enter plants primarily in three compounds: CO_2, H_2O, and O_2. All other elements are minerals present in ionized form in the soil, and they enter plants dissolved in water. Nitrogen, sulfur, and phosphorus are present in the soil largely as anions (negatively charged ions): nitrates (NO_3^-), sulfates (SO_4^{--}), and phosphates (PO_4^{---}). The remaining elements are present as cations (positively charged ions): K^+, Ca^{++}, Mg^{++}, Fe^{+++} (or Fe^{++}). Nitrogen is also present in the ammonium ion (NH_4^+). Trace elements (MICRONUTRIENTS) are required in only minute quantities, but they are essential for plants. Boron, copper, manganese, zinc, and molybdenum are known to be micronutrient elements.

SOIL is the natural environment of most roots. It is composed of solid mineral particles, organic substances, air, water, dissolved minerals, and soil organisms. The characteristics and utility of a particular soil depend on the proportions of these constituents.

The SOLID MINERAL PARTICLES are classified as SAND (particles 0.02 to 2.0 mm. diameter), SILT (particles 0.002 to 0.02 mm. diameter), and CLAY (particles having diameters less than 0.002 mm.). LOAMS are mixtures of sand, silt, and clay.

ORGANIC SUBSTANCES range from intact bodies of plants or animals that have recently died to HUMUS (partially decomposed plant and animal materials) to the soluble, simple organic compounds that result from complete decomposition. A certain amount of humus is desirable for it binds clay or silt particles into larger aggregates.

The PORE SPACES between soil particles contain AIR OR WATER. Under

(Continued on page 36)

ADDITIONAL INFORMATION

CARBON, HYDROGEN, and OXYGEN are important components of nearly all organic compounds. Free oxygen is necessary for aerobic respiration. NITROGEN is essential for plants because it is present in amino acids, proteins (and therefore enzymes), and some vitamins. Both nitrogen and PHOSPHORUS are constituents of ADP, ATP, NAD, NADP, CoA, and nucleic acids. SULFUR is present in three amino acids, proteins, CoA, and a few vitamins. CALCIUM plays at least two roles in plants: it is part of the middle lamella of the cell wall, and it affects the permeability of membranes. MAGNESIUM is part of the chlorophyll molecule, and IRON is part of several enzymes, including the cytochromes. Iron is also necessary for the synthesis of chlorophyll. Magnesium or iron deficiency may result in CHLOROSIS. The role of POTASSIUM is unknown; it is the only major element in plants that does not become part of an organic compound. Trace elements are required in such small quantities that they probably have catalytic functions. Mineral deficiencies may result in STUNTED GROWTH and in NECROSIS, which is the death of certain parts of a plant.

Plants absorb not only those elements that they require, but also substances that they do not need and which may be toxic. An excess of one element often interferes with the functioning of another. Trace elements, which are essential in small amounts, sometimes accumulate until they become toxic.

Soil water is classified as either gravitational water, capillary water, or hygroscopic water. GRAVITATIONAL WATER is that water which soil cannot hold against gravity. It remains in the soil only temporarily after a flood or rain; it moves downward to the water table or laterally to drier soil, and in so doing it may leach out soluble minerals. Plants can absorb gravitational water while it remains in the soil. CAPILLARY WATER is that water which is held against gravity and is available to plants. It fills small pore spaces and adheres to the surfaces of soil particles. HYGROSCOPIC WATER is soil water that is unavailable to plants because it is tenaciously held by soil particles.

The FIELD CAPACITY of a soil is the percentage of water that it can hold against gravity (its capillary and hygroscopic water). The field capacity of a clay soil is high, that of a sandy soil is low. The PERMANENT WILTING PERCENTAGE of a soil is the percentage of water it holds when plants growing in it reach permanent wilting. (A plant is permanently wilted if it has removed all available water from the soil and does not recover when placed in a dark, moist chamber.) The permanent wilting percentage of a clay soil is high, that of a sandy soil is low. The difference between field capacity and permanent wilting percentage is the STORAGE CAPACITY of a soil.

(Continued on page 36)

EXPLANATIONS

1. *Some minerals known to be required in large amounts for plant growth are* phosphorus, potassium, sulfur, calcium. The ten major essential elements for plants can be easily remembered by the mnemonic device in which their atomic symbols are arranged thus: C HOPKNS CaFe Mg (C. Hopk'ns cafe, mighty good).

2. *Magnesium is required by plants because it is part of* chlorophyll molecules. It also has other catalytic functions.

3. *Soil particles arranged in order of increasing size are* clay, silt, and sand. They are also arranged in decreasing order of ability to hold water.

4. *Clay soils are characterized by* small pore spaces and high field capacity. Clay particles are of colloidal size and tend not only to retain water in the small pore spaces among them, but also to imbibe water.

5. *A fertile soil is likely to have a pH of* 6 to 7. Most fertile soils are neutral or slightly acid. Soils with a pH as low as 3 or as high as 11 are rare. Few plants can live at such pH extremes.

6. *Water that remains in the soil only temporarily after a rain* is gravitational water. This is the water that the soil is unable to hold against gravity.

7. *Soil water that is available to plants is* gravitational water and capillary water. Hygroscopic water is so strongly held by the soil that plants cannot absorb it.

8. *If the concentration of an element is higher in a plant than it is in the surrounding soil we may assume that* the plant expends energy when the element enters. The movement of a substance against a diffusion pressure gradient always requires the expenditure of energy.

9. *A solution with a pH of 2 has a concentration of hydrogen ions that is* one hundred times that of a solution with a pH of 4. It is ten times that of a solution with a pH of 3, which in turn is ten times that of a solution with a pH of 4.

10. *Components of soils that are required by plants include* water, air, and dissolved minerals. That solid particles are not necessary for plants is shown by the fact that plants can be raised successfully in aerated mineral solutions that contain no solid matter.

Answers

d	1
a	2
b	3
b	4
c	5
a	6
b	7
d	8
d	9
c	10

ordinary conditions large spaces (macropores) hold mostly air, and small spaces (micropores) hold water. Since air provides the oxygen required by roots, and since soil water is almost the only source of water for plants, a combination of large and small spaces in soil is most desirable. The anions of DISSOLVED MINERALS are free in the soil water; the cations are adsorbed on colloidal soil particles.

SOIL ORGANISMS are both micro- and macroscopic. They include protozoa, algae, fungi, bacteria, worms, and insects. Most of them decompose dead organisms. They digest macromolecules to simple organic compounds which are then converted to water, gases, and minerals. Soil organisms affect the characteristics of soils, but soil characteristics also determine what organisms can live in each type of soil.

The ENTRANCE OF MINERALS INTO PLANTS is usually not by simple diffusion, for plants tend to accumulate some minerals until the concentrations of these minerals are higher in the cells than they are in the soil. This means that in entering the plant the minerals move against a diffusion pressure gradient, that is, in a direction opposite to that in which they would move by diffusion. Energy must be supplied whenever materials move against a diffusion pressure gradient. Evidence is strong that plants use respiratory energy (high-energy bonds of ATP) in pumping minerals into the root. Once absorbed, minerals are usually transported upward in the xylem, but they may also be carried in the phloem. Those moving in the phloem usually have become part of some organic compound.

Soils are depleted of minerals by floods, which leach away soluble minerals, and by the repeated removal of crops. Minerals may be replaced by application of FERTILIZERS. Chemical fertilizers can be obtained with different proportions of minerals suited to the requirements of a variety of soils. They usually contain nitrogen, phosphorus (as phosphoric acid), and potassium (as potash); also some lime to reduce acidity. Organic fertilizers consist of decomposing plant and animal parts and waste products; they do not always have the desired balance of minerals, but they do improve soil texture and fertility. Application of too much fertilizer may cause plasmolysis of root cells and death of plants.

Most soils are mixtures of sand, silt, clay, and humus, these components being present in varying amounts. Sandy soils have large pore spaces and are well aerated, but they have a very low water-holding capacity and tend to be easily blown by wind when dry. Clay soils have small pore spaces and such high water-holding capacity that air often is excluded from the pores and the soil becomes anaerobic. A good balance between soil particles of various sizes is needed by many plants, though there are some plants adapted to growing in almost any type of soil.

The ACIDITY OF ALKALINITY of a soil is expressed as its pH. An acid solution is one that has more hydrogen ions (H^+) than hydroxyl ions (OH^-), whereas an alkaline (or basic) solution has more hydroxyl ions than hydrogen ions. A neutral solution has equal numbers of the two ions. A pH of 7 indicates neutrality; a pH numerically lower than 7 indicates acidity; and a pH higher than 7 indicates alkalinity.

The pH scale is a logarithmic one. Consecutive whole numbers differ by factors of 10 with regard to ion concentration. For example, a solution with a pH of 6 has a H^+ concentration ten times higher than that of a solution with a pH of 7, but its OH^- concentration is one-tenth that of a solution with a pH of 7.

Most plants thrive best in soils with a pH of 6 or 7. Few plants can survive or grow well at a pH above 9 or below 4. Acid soils are usually not fertile because hydrogen ions replace the mineral cations that are normally located on colloidal soil particles. When the cations go into solution they are easily lost by leaching. Acid soils are frequently treated with lime (calcium hydroxide or calcium carbonate), for the calcium ions replace hydrogen ions on the soil particles.

10 STEMS

SELF-TEST

1. The primary functions of stems are
   a storage and support
   b support and conduction
   c conduction and storage

2. Branch primordia and leaf primordia are produced from the
   a apical meristem
   b pericycle
   c vascular cambium
   d cortex

3. The primary tissues of most stems are
   a epidermis, cortex, xylem, phloem, and pith
   b epidermis, endodermis, pericycle, xylem, and phloem
   c pericycle, cortex, xylem, and phloem
   d epidermis, cortex, endodermis, and pericycle

4. In the primary tissues of the stem, the cambium separating the xylem and phloem is called
   a procambium
   b fascicular cambium
   c cork cambium
   d interfascicular cambium

5. Monocot stems differ from dicot stems in having
   a scattered vascular bundles and secondary growth
   b vascular bundles arranged in a ring and secondary growth
   c scattered vascular bundles and no secondary growth
   d vascular bundles arranged in a ring and no secondary growth

6. The cells that transport materials vertically in stems are
   a xylem and phloem ray cells
   b tracheids, vessel elements, and sieve-tube elements
   c tracheids, ray cells, and fibers
   d xylem and phloem fibers

7. Secondary xylem consists of
   a only living cells
   b only dead cells
   c both living and dead cells

8. A growth ring in wood
   a consists of the spring wood and summer wood of the same year
   b consists of the summer wood of one year and the spring wood of the following year
   c is the junction between spring wood and summer wood of the same year
   d is the junction between summer wood of one year and spring wood of the following year

9. Gaseous interchange between the air and the internal tissues of older, corky stems takes place through
   a sieve plates
   b pits
   c lenticels
   d stomata

10. The mechanism of food transport in the phloem is
   a diffusion
   b cytoplasmic streaming
   c mass flow
   d not understood

11. Onions are
   a tubers
   b rhizomes
   c bulbs
   d corms

1 ______
2 ______
3 ______
4 ______
5 ______
6 ______
7 ______
8 ______
9 ______
10 ______
11 ______

BASIC FACTS

Stems are typically aerial organs with the functions of supporting other aerial organs and conducting water and dissolved materials from one organ to another. Stems are either WOODY or HERBACEOUS.

PRIMARY GROWTH of stems is similar to that of roots. At the tip of each stem is an APICAL MERISTEM which lies within the terminal bud. Below this are REGIONS OF ELONGATION AND MATURATION which are longer and tend to overlap more than they do in roots. LEAF PRIMORDIA (young leaves) arise from the surface of the apical meristem a short distance below its tip. One BRANCH PRIMORDIUM arises in the AXIL (the angle formed by leaf and stem) of each leaf primordium. Each branch primordium develops into a LATERAL (or AXILLARY) BUD with its own stem tip and leaf primordia. In woody plants the stem tip is protected by BUD SCALES. When buds of perennial plants open in the spring, the bud scales abscise, leaving a ring of BUD SCALE SCARS. The short length of stem within the bud elongates and the leaves expand. During the growing season a new terminal bud is formed which remains dormant over winter and opens the following spring. One ring of bud scale scars is formed on a twig each year.

Three ARRANGEMENTS OF LEAVES ON STEMS are possible: ALTERNATE (one leaf at a node), OPPOSITE (two leaves at opposite sides of a node), and WHORLED (three or more leaves symmetrically arranged at a node). When leaves abscise, leaf scars are left on the surface of the stem. These are covered with a layer of cork that heals the scars.

The PRIMARY TISSUES of a DICOT STEM are epidermis, cortex, primary xylem, primary phloem, and pith. Some young stems have a photosynthetic cortex and have stomata in the epidermis. Vascular bundles are ar-

(Continued on page 40)

ADDITIONAL INFORMATION

DICOT WOODS usually consist of five types of XYLEM CELLS: tracheids, vessel elements, xylem fibers, xylem parenchyma cells, and xylem ray cells. Gymnosperm woods generally have only tracheids and ray cells. TRACHEIDS are vertically elongated cells that taper at both ends. VESSEL ELEMENTS usually have a larger diameter than do tracheids. At maturity they have no end walls and are arranged end to end in vertical rows forming continuous tubes called vessels. Tracheids and vessels have no protoplasm at maturity. Their secondary cell walls show various patterns: annular, spiral, scalariform, reticulate, and pitted. XYLEM FIBERS are generally longer and thinner than tracheids and may have thicker secondary cell walls. XYLEM PARENCHYMA CELLS and XYLEM RAY CELLS are living. The former are vertically elongated; the latter are horizontally elongated and form the xylem rays which lie along the radii of the stem.

The xylem cells produced in spring are generally large; those produced in summer are smaller and usually have thicker secondary cell walls. The difference between spring and summer wood makes the growth rings obvious. RING-POROUS WOODS are those in which vessels are larger and more numerous in spring wood; this accentuates the difference between spring and summer wood. DIFFUSE-POROUS WOODS have vessels uniformly distributed in both spring and summer wood. The age of any segment of a woody stem can be estimated by counting growth rings. Counts are not always accurate because two rings may be produced in some years.

As secondary xylem becomes older, the tracheids and vessels cease to function and the living cells (parenchyma and rays) die. This nonfunctioning wood is called HEARTWOOD, and it occupies the center of old woody stems. The younger, functioning wood through which water rises is SAPWOOD, which lies between heartwood and cambium. Heartwood is generally darker than sapwood.

DICOTS have five types of PHLOEM CELLS: sieve-tube elements, companion cells, phloem fibers, phloem parenchyma cells, and phloem ray cells. Gymnosperm phloem generally has only SIEVE CELLS (rather than sieve-tube elements), phloem parenchyma, and phloem rays. SIEVE-TUBE ELEMENTS are living, thin-walled, vertically elongated cells placed end to end in vertical rows called SIEVE TUBES. The end walls are pitted and are called SIEVE PLATES. The cytoplasm of adjacent sieve-tube elements is connected by plasmodesmata that run through the pits. Sieve-tube elements are peculiar in having cytoplasm but no nuclei at maturity. Adjacent to each sieve-tube element is at least one COMPANION CELL which has a nucleus that is presumed to serve both cells. Movement of food occurs in the sieve tubes. FIBERS, PARENCHYMA CELLS, and RAY CELLS in phloem resemble those in xylem.

(Continued on page 40)

EXPLANATIONS

1. *The primary functions of stems are* support of leaves, flowers, and fruits and conduction of materials among these organs and roots.

2. *Branch primordia and leaf primordia are produced from the* apical meristem. This is in contrast to branch root primordia which develop from an internal tissue (pericycle).

3. *The primary tissues of most stems are* epidermis, cortex, xylem, phloem, and pith. Endodermis and pericycle are not present in most stems.

4. *In the primary tissues of the stem, the cambium separating the xylem and phloem is called* fascicular cambium. Interfascicular cambium lies between the bundles. Fascicular and interfascicular cambium make a continuous ring of tissue.

5. *Monocot stems differ from dicot stems in having* scattered vascular bundles and no secondary growth. Not all dicot stems have secondary growth, but even herbaceous dicots frequently have some small amount of cambial activity.

6. *The cells that transport materials vertically in stems are* tracheids and vessel elements, which conduct water, and sieve-tube elements, which conduct food. Xylem and phloem rays transport materials laterally.

7. *Secondary xylem consists of* both living and dead cells. Tracheids, vessel elements, and xylem fibers are dead at maturity. Xylem parenchyma cells and xylem ray cells are living. Secondary phloem consists of both living and dead cells. Sieve-tube elements, companion cells, phloem parenchyma cells, and phloem ray cells are living. Unlike those of xylem, conducting cells in phloem (sieve-tube elements) are alive. Phloem fibers are dead at maturity.

8. *A growth ring in wood* consists of the spring wood and summer wood of the same year. Many large xylem cells are produced in spring; at other seasons a few small xylem cells are formed. The junction between the summer wood of one year and the spring wood of the following year makes the rings readily visible.

9. *Gaseous interchange between the air and the internal tissues of older, corky stems takes place through* lenticels. Lenticels permit entrance of oxygen to and exit of carbon dioxide from living stem tissues.

10. *The mechanism of food transport in the phloem is* not understood. No hypothesis thus far advanced adequately explains how phloem conducts food at the rate it does.

11. *Onions are* bulbs; they consist of short, vertical, underground stems with fleshy, circular leaves.

Answers

b	1
a	2
a	3
b	4
c	5
b	6
c	7
a	8
c	9
d	10
c	11

Origin of Tissues of the Stem

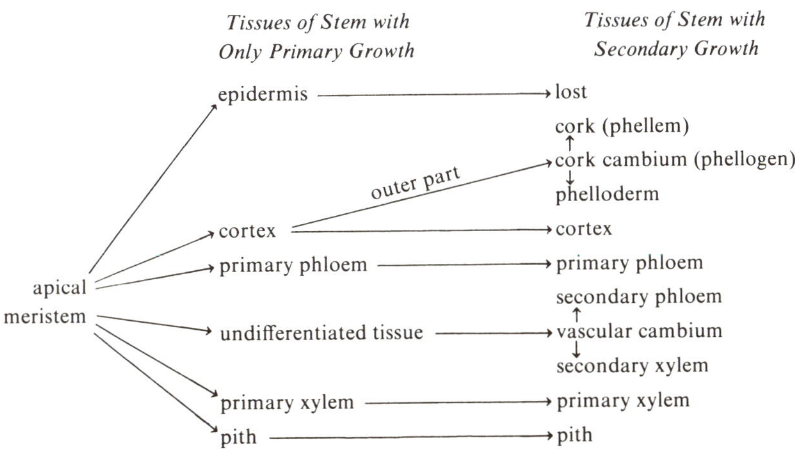

ranged in a ring between cortex and pith. Usually primary xylem occupies the inner part of each vascular bundle and primary phloem the outer part, although other arrangements do occur in some species. It is not uncommon for pith cells to disintegrate, leaving a hollow center. MONOCOT STEMS have the same primary tissues as dicot stems, but numerous vascular bundles are scattered throughout the stem.

SECONDARY GROWTH of stems resembles that of roots in that vascular cambium produces secondary xylem and secondary phloem, and cork cambium produces cork and phelloderm. The VASCULAR CAMBIUM which arises from undifferentiated tissue lying between the xylem and phloem of the vascular bundles is called FASCICULAR CAMBIUM. INTERFASCICULAR CAMBIUM originates between the bundles and makes the vascular cambium one continuous tissue. Activity of vascular cambium is greatest in spring; it declines during summer and no new secondary tissues are formed in winter. This pattern of activity produces the GROWTH RINGS in secondary xylem. Such rings occur in secondary phloem but they are not so conspicuous as in xylem.

CORK CAMBIUM most often arises from the outermost cortical cells, but it may be produced by inner cortical cells or even by phloem. When cork cambium begins its activity, any tissues external to it are lost. Cork is impermeable to oxygen, but it has structures called LENTICELS which are channels of gaseous exchange between the air and the internal respiring, nonphotosynthetic stem tissue.

WOOD is secondary xylem. BARK consists of all tissues external to the vascular cambium.

FOOD TRANSLOCATION occurs mainly in the phloem. Foods move both upward and downward in the sieve tubes, and different substances may travel in opposite directions simultaneously. Glucose is ordinarily converted to the disaccharide sucrose before transport. The rate of sucrose movement in the phloem has been calculated to be more than 100 cm. per hour; this is too rapid to be explained by diffusion. It is known that lowering the temperature or reducing the oxygen supply will slow the transport of food, but there is at present no explanation of the mechanism of food transport. One hypothesis has been advanced that the STREAMING OF CYTOPLASM in sieve-tube elements carries food from one end of the cell to the other end where it passes through the sieve plate to the next sieve-tube element; such a mechanism could easily explain how materials might move in opposite directions in the phloem, but cytoplasmic streaming has never been satisfactorily demonstrated in mature sieve tubes. The MASS FLOW hypothesis suggests that a high turgor pressure in the leaf forces the sucrose solution to flow through the phloem to the root or to some other organ with a low turgor pressure. This hypothesis is supported by the fact that sap continues to exude for some time from cut phloem, indicating that the sap is under pressure, but the hypothesis cannot explain the simultaneous movement of materials in opposite directions in the phloem. Neither hypothesis can explain the rapid rate of food translocation.

STEM MODIFICATIONS assume a number of forms. RHIZOMES, also called ROOTSTOCKS (e.g., iris), are perennial, horizontal underground stems that send up new shoots each year. Rhizomes usually serve as food-storage organs, and their underground location protects the plant from fire and from grazing animals. TUBERS (e.g., potato) are enlarged, fleshy portions of rhizomes. They are usually rich in stored food. BULBS (e.g., onion) contain short, vertical underground stems surrounded by large, fleshy leaves. CORMS (e.g., gladiolus) superficially resemble bulbs, but they are hard stems covered by several papery scale leaves. STOLONS, or RUNNERS (e.g., strawberry plant), are horizontal stems that grow just above the surface of the ground. New shoots are formed at nodes that touch the soil. Some THORNS and TENDRILS are modified stems. Stems of many cactus plants are both PHOTOSYNTHETIC and WATER-STORING organs. CLADOPHYLLS are leaflike stems. ASEXUAL REPRODUCTION by stems is common. Many, but not all, stems will form new roots if they are removed from the plant and kept moist.

11 FLOWERS

SELF-TEST

1. Beginning at the outside and proceeding toward the center of the flower, the order in which flower parts occur is
   a sepals, petals, stamens, and pistils
   b petals, sepals, stamens, and pistils
   c sepals, petals, pistils, and stamens
   d petals, sepals, pistils, and stamens

2. The essential parts of a flower are
   a sepals and petals
   b petals and stamens
   c stamens and pistils
   d pistils and sepals

3. An imperfect flower is one that lacks either
   a sepals or petals
   b petals or stamens
   c stamens or pistils
   d pistils or sepals

4. Dioecious species are those which have
   a staminate and pistillate flowers on the same plant
   b staminate and pistillate flowers on separate plants
   c perfect flowers on all plants
   d perfect flowers and imperfect flowers on separate plants

5. A mature embryo sac usually consists of
   a one cell
   b two cells
   c five cells
   d seven cells

6. Pollination is the
   a union of a sperm nucleus with an egg nucleus
   b transfer of pollen grains from an anther to a stigma
   c growth of a pollen tube down a style
   d penetration of an embryo sac by a pollen tube

7. Double fertilization is the fusion of
   a two sperm nuclei with one egg nucleus
   b a sperm nucleus with two egg nuclei
   c a sperm nucleus with an egg nucleus and another sperm nucleus with two polar nuclei
   d a sperm nucleus with two egg nuclei and another sperm nucleus with a polar nucleus

8. A microspore mother cell forms
   a pollen grains
   b embryo sacs
   c a pollen sac
   d an ovule

9. An inflorescence is a(n)
   a fusion of floral parts
   b unopened flower bud
   c infertile flower
   d cluster of flowers

10. A pedicel is
   a the stalk that bears the anther
   b the stalk that bears the stigma
   c a stem on which an inflorescence or a flower is borne
   d a stem on which an individual flower of an inflorescence is borne

1 ______
2 ______
3 ______
4 ______
5 ______
6 ______
7 ______
8 ______
9 ______
10 ______

BASIC FACTS

The function of the flower is sexual reproduction. A flower consists of four kinds of parts: sepals, petals, stamens, and pistils. A pistil consists of one or more carpels. The parts are located on the RECEPTACLE, the tip of a flower-bearing stem.

The outermost whorl of flower parts is the CALYX, consisting of several sepals. SEPALS commonly are green and shaped like simple, entire leaves. Before the flower opens, they provide protection to the inner parts. To the inside of the calyx is the COROLLA, a whorl of petals. PETALS are leaflike in shape, but often are brightly colored. The calyx and corolla together are the PERIANTH. If sepals and petals are alike, they are called TEPALS.

To the inside of the corolla are one or more whorls of stamens. Each STAMEN consists of a FILAMENT, or stalk, that bears at its top an ANTHER, within which POLLEN is formed. All the stamens of a flower comprise the ANDROECIUM. Occupying the center of a flower are one or more CARPELS. A simple PISTIL consists of one carpel, a compound pistil of two or more carpels fused together. All the pistils of a flower comprise the GYNOECIUM. Each pistil has a somewhat swollen lower portion, the OVARY, which narrows into a slender STYLE topped by a STIGMA, the portion receptive to pollen.

An ovary may have several cavities, or LOCULES, separated by partitions. Each ovary contains one or more ovules, which, after fertilization, ripen into seeds. The portion of the ovary that bears the ovules is the PLACENTA. Each ovule has its own short stalk called a FUNICULUS. When ready for fertilization, each ovule contains an EMBRYO SAC surrounded by a single layer of cells, the NUCELLUS, and by one or two INTEGUMENTS. A small opening, the MICROPYLE, exists in the integuments. A mature EMBRYO SAC usually consists of seven cells: one EGG cell, two SYNERGID cells, three

(Continued on page 44)

ADDITIONAL INFORMATION

The four flower parts are considered by many botanists to be modified leaves. The leaflike shapes of SEPALS and PETALS are obvious. CARPELS are thought to be modified leaves which bear ovules on their margins and which have folded, the two margins meeting and fusing. The individual CARPELS of a flower may remain as separate pistils or fuse into a single compound pistil. STAMENS are so greatly modified that their leaflike nature is not obvious, but in some flowers, structures appear that are intermediate in form between petals and stamens.

COALESCENCE, or the fusion of certain floral parts, is an evolutionary development exhibited by a number of plants. Petals may fuse to form a COROLLA TUBE; sepals also may be joined, as may stamens and carpels. The bases of the sepals, petals, and stamens may be united into a FLORAL TUBE; in other plants the floral tube in turn has fused with the ovary, making the latter appear to occupy an INFERIOR position.

A PEDUNCLE is the stem on which an INFLORESCENCE or an isolated flower is borne. A PEDICEL is the short stalk on which an individual flower of an inflorescence is borne. Some common types of inflorescence are: raceme, corymb, spike, catkin, panicle, umbel, and head. A RACEME consists of a single peduncle along the length of which occur the flowers, each on its own pedicel; the pedicels are of nearly the same length and are evenly spaced along the peduncle. A CORYMB resembles a raceme, but the pedicels of the lower flowers are longer than those of the upper flowers, making the top of the inflorescence flat or rounded. A SPIKE is similar to a raceme, but the individual flowers are sessile, that is, they have no pedicels. Spikes consisting of imperfect flowers are CATKINS. A PANICLE is a branched raceme. An UMBEL has pedicels which all arise from the tip of the peduncle; this produces a rounded or flat-topped inflorescence. In a HEAD the individual flowers are sessile or have very short pedicels and are so closely crowded on a short, sometimes flattened peduncle that they often give the appearance of being a single flower. A CYME is an inflorescence in which the terminal flower opens first; the youngest flowers are near the base of the inflorescence.

Within each young OVULE is a single MEGASPORE MOTHER CELL (EMBRYO SAC MOTHER CELL). This cell like nearly all other cells of the plant is DIPLOID, that is, all the chromosomes in its nucleus are represented as pairs. The megaspore mother cell undergoes MEIOSIS, a set of two nuclear divisions resulting in the formation of four cells called MEGASPORES. These cells are HAPLOID; they contain half the number of chromosomes of diploid cells, and only one member of each pair is represented. In most flowering plants three of the megaspores disintegrate, and the nucleus of the remaining megaspore undergoes three mitotic divisions, pro-

(Continued on page 44)

EXPLANATIONS

1. *Beginning at the outside and proceeding toward the center of the flower, the order in which flower parts occur is* sepals, petals, stamens, and pistils. Many variations occur in flowers, but the position of these four parts relative to each other is constant (except in those cases in which some parts are missing).

2. *The essential parts of a flower are* stamens and pistils. They are called essential because they produce the sperm and egg nuclei. Sepals and petals are not immediately involved with reproduction although if brightly colored they may attract insects that are normally required for pollination. Sepals protect the other flower parts before the flower buds open.

3. *An imperfect flower is one that lacks either* stamens or pistils. An incomplete flower lacks one or more of the four main floral parts.

4. *Dioecious species are those which have* staminate and pistillate flowers on separate plants. The word dioecious comes from the Greek and means "two houses."

5. *A mature embryo sac usually consists of* seven cells. It begins as a one-nucleate cell, and repeated mitotic divisions produce two, four, and finally eight nuclei. Cell walls form, dividing this cell into seven cells, one of which has two nuclei.

6. *Pollination is the* transfer of pollen grains from an anther to a stigma. Pollination should not be confused with fertilization, which is the union of a sperm nucleus with an egg nucleus. In flowering plants pollination must precede fertilization; however pollination is not invariably followed by fertilization, for pollen grains may fail to germinate on the stigma, or the pollen tube may not reach or penetrate an ovule.

7. *Double fertilization is the fusion of* a sperm nucleus with an egg nucleus and another sperm nucleus with two polar nuclei. The first fusion results in the formation of a zygote, the latter in the formation of the endosperm. Double fertilization is peculiar to flowering plants.

8. *A microspore mother cell forms* pollen grains. Each microspore mother cell divides into four microspores, each of which matures into a pollen grain.

9. *An inflorescence is a* cluster of flowers arising from the same peduncle.

10. *A pedicel is* a stem on which an individual flower of an inflorescence is borne. The stem on which the entire inflorescence is borne is called a peduncle.

Answers

Answer	No.
a	1
c	2
c	3
b	4
d	5
b	6
c	7
a	8
d	9
d	10

ANTIPODAL cells, and one cell with two POLAR NUCLEI.

POLLINATION is the transfer of pollen from anthers to stigmas. When pollen lands on a receptive stigma it germinates there, sending a POLLEN TUBE down the style to the ovary. The pollen tube enters an ovule, usually through the micropyle, and discharges its two SPERM NUCLEI into the embryo sac. One nucleus fertilizes the egg nucleus, the other the two polar nuclei (DOUBLE FERTILIZATION). The fertilized egg is a ZYGOTE, which develops into an EMBRYO; the fertilized polar nuclei develop into the ENDOSPERM, a tissue that stores food to be used by the germinating embryo. As these events occur, the ovule ripens into a SEED, and the ovary ripens into a FRUIT.

Not all flower parts are present in all flowers. In some species either sepals, petals, or both may be lacking, or stamens and pistils may be present in different flowers. Flowers with all four parts are COMPLETE; those which do not have all four are INCOMPLETE. If both stamens and pistils are present, the flowers are called PERFECT; if either stamens or pistils are lacking, the flowers are IMPERFECT. In MONOECIOUS species staminate and pistillate flowers occur on the same plant; in DIOECIOUS species they occur on separate plants.

A cluster of more or less tightly grouped flowers is an INFLORESCENCE. BRACTS are modified leaves that may appear below an inflorescence. Usually they are small and inconspicuous, but they may be large and showy as they are in poinsettia.

ducing an eight-nucleate cell, the mature EMBRYO SAC. All eight nuclei are haploid. The nuclei distribute themselves within the embryo sac so that one egg nucleus and two synergid nuclei are near the micropylar end, three antipodal nuclei are at the opposite end, and two polar nuclei are in the center. Frequently cell walls form around egg, synergid, and antipodal nuclei, making the embryo sac seven-celled.

Each ANTHER usually contains two POLLEN SACS, each of which is separated into two POLLEN CHAMBERS. Each pollen chamber is lined by a nutritive tissue, the TAPETUM. Within the pollen sacs are many diploid MICROSPORE MOTHER CELLS (POLLEN MOTHER CELLS), each of which undergoes meiosis, forming a tetrad of haploid MICROSPORES. All four of these microspores are functional. They separate from each other, and each undergoes a mitotic division, becoming a two-celled POLLEN GRAIN. One cell is the TUBE CELL, the other the GENERATIVE CELL. Pollen grains are usually shed in the two-celled stage. The wall between the two pollen chambers of each pollen sac breaks down, and a slit develops through which the pollen grains escape. When a pollen grain lands on a receptive stigma, a POLLEN TUBE develops from the pollen grain and grows down the style to the ovary. The generative nucleus moves into the pollen tube and undergoes a mitotic division that produces two haploid SPERM NUCLEI. The tube nucleus also moves into the pollen tube but usually disintegrates there.

When one haploid sperm nucleus fertilizes the haploid egg nucleus, the diploid condition is restored in the zygote. The fusion of the other sperm nucleus with two haploid polar nuclei results in the formation of a triploid ENDOSPERM NUCLEUS (TRIPLE FUSION NUCLEUS). The endosperm of flowering plants is a unique tissue in that it is normally POLYPLOID.

CROSS-POLLINATION is the transfer of pollen from the stamens of one plant to a pistil of another plant. SELF-POLLINATION is the transfer of pollen from the stamens to a pistil of the same plant. Pollen of WIND-POLLINATED flowers is light and dry; the perianth parts are reduced in size and thus do not interfere with the movement of pollen. Pollen of INSECT-POLLINATED flowers is heavy and sticky and clings readily to the bodies of insects; the perianth parts tend to be large and brightly colored. NECTARIES (nectar-producing tissues) are also common among insect-pollinated flowers. A few species are adapted to pollination by birds and other animals or by water.

12 SEEDS AND FRUITS

SELF-TEST

1. The seed of a flowering plant is a ripened
   a embryo sac
   b pistil
   c ovary
   d ovule

2. When the bean seed germinates, the first part to break through the soil is the
   a cotyledon
   b epicotyl
   c hypocotyl
   d coleoptile

3. The radicle is
   a found at the tip of the hypocotyl
   b found at the tip of the epicotyl
   c a hard outer layer of the pericarp
   d a small opening in the seed coat

4. The general function of the cotyledon is to supply the embryo with
   a water
   b minerals
   c food
   d oxygen

5. The coleorhiza protects the
   a leaf primordia
   b stem tip
   c cotyledons
   d radicle

6. The first part to emerge from a germinating seed is the
   a hypocotyl
   b radicle
   c leaf primordium
   d stem tip

7. Afterripening of a seed occurs when
   a sufficient water is present in the soil
   b the seed coat splits or dissolves
   c exposure to light is excessive
   d the seed undergoes physiological changes

8. In the strictest sense of the term a fruit is a ripened
   a embryo sac
   b pistil
   c ovary
   d ovule

9. A drupe is a
   a fleshy, one-seeded fruit with a hard endocarp
   b fleshy, many-seeded fruit derived from a compound pistil
   c dry, one-seeded fruit, derived from a simple pistil
   d dry, many-seeded fruit derived from a compound pistil

10. Dry fruits are
    a all dehiscent
    b all indehiscent
    c either dehiscent or indehiscent

11. A strawberry is a(n)
    a simple fruit
    b simple accessory fruit
    c aggregate fruit
    d aggregate-accessory fruit

12. Wind-disseminated seeds and fruits tend to have
    a barbs or hooks
    b hairs or wings
    c sticky surfaces
    d rounded, compact shapes

1 ________
2 ________
3 ________
4 ________
5 ________
6 ________
7 ________
8 ________
9 ________
10 ________
11 ________
12 ________

BASIC FACTS

The parts of a SEED commonly are embryo, endosperm, and seed coat. After double fertilization occurs, the ovule ripens into a seed. The zygote forms an EMBRYO, the endosperm cell develops into the ENDOSPERM tissue, and the integuments become the SEED COATS. In most seeds the nucellus disintegrates, but in a few species it remains as the PERISPERM.

A mature EMBRYO consists of an axis bearing one or more COTYLEDONS, or seed leaves. Monocot embryos have one cotyledon, dicot embryos have two, and gymnosperm embryos have several. The portion of the axis above the point of attachment of the cotyledons is the EPICOTYL, or PLUMULE; the portion below is the HYPOCOTYL. The epicotyl consists of a stem tip and its leaf primordia. The lower part of the hypocotyl terminates in a RADICLE which gives rise to the primary root. In corn and other grasses the epicotyl is protected by a sheath, the COLEOPTILE, and the radicle by another sheath, the COLEORHIZA.

Three external characteristics seen on many seeds are: the MICROPYLE, a small round opening in the seed coat; the HILUM, the scar left by the breaking off of the funiculus; and the RAPHE, a ridge formed by the fusion of the base of the funiculus with the seed coats.

A FRUIT is a ripened ovary. While seed development occurs, the ovary enlarges into a fruit, its wall thickening into the PERICARP. The pericarp has three layers: EXOCARP, MESOCARP, and ENDOCARP, which are outer, middle, and inner layers, respectively. Sepals, petals, stamens, stigmas, and styles often drop while the fruit is ripening, but it is not uncommon for several of these flower parts to remain on the fruit.

There are three main types of fruits: simple, aggregate, and multiple. A SIMPLE FRUIT is derived from the ovary of a flower with a single pistil. An AG-

(Continued on page 48)

ADDITIONAL INFORMATION

All the cells of the EMBRYO are derived from the zygote by mitotic divisions. The first divisions produce a PROEMBRYO, a single row of cells. The cell that is deepest in the embryo sac develops into the embryo proper; the remaining cells comprise the SUSPENSOR. While the embryo develops, the endosperm nucleus undergoes mitotic divisions, producing a free-nuclear endosperm that is a single, large, multinucleate cell. At a later stage, the endosperm becomes cellular as cell walls develop around the nuclei. The suspensor pushes the developing embryo into the endosperm.

When seeds GERMINATE they imbibe water, the protoplasm becomes hydrated, and chemical activities, especially respiration, are resumed at a high rate. The radicle is the first organ to penetrate the seed coat; it then forms the primary root. The stem elongates and its leaves expand as the epicotyl grows out of the seed. These processes use energy from the food stored in cotyledons or endosperm. In some species, such as CORN, the endosperm persists in the seed and becomes the principal food-storage tissue. The main function of the cotyledon (called a SCUTELLUM in corn) is digestion and absorption of food from the endosperm; it does not emerge from the seed, but remains in contact with the endosperm. In the CASTOR BEAN the cotyledons are thin and leaflike in shape. When germination begins, they absorb food from the endosperm, but later they emerge from the seed, expand, and become green and photosynthetic. In other seeds, like BEANS, the endosperm disintegrates early and food is stored mainly in the cotyledons. In a few species, such as CARNATION, there is no endosperm, but food is stored in the PERISPERM.

When BEAN seeds germinate, the primary root emerges from the seed and develops root hairs and, later, branch roots. As the root becomes established in the soil, the hypocotyl increases in length and becomes arched. The arch emerges and straightens when exposed to light, raising the cotyledons and epicotyl above the soil. The cotyledons soon expand, tearing the seed coat, which drops off. The stem elongates, its young leaves expand, and new leaves form. As more and more food drains from the cotyledons they shrivel and fall. The new vegetative leaves are capable of photosynthesizing sufficient food for the plant.

When the primary root of CORN emerges from the kernel it is sheathed by its coleorhiza. Shortly after emergence, the coleorhiza ceases to grow and the root tip breaks through it. In similar fashion the coleoptile protects the epicotyl during emergence from the kernel and growth upward through the soil. The epicotyl then breaks through the coleoptile. The primary root of corn dies early. The root system is entirely adventitious, roots forming at successively higher levels on the stem.

(Continued on page 48)

EXPLANATIONS

1. *The seed of a flowering plant is a ripened* ovule. An ovule develops into a seed only after fertilization.

2. *When the bean seed germinates, the first part to break through the soil is the* hypocotyl. As the hypocotyl elongates it raises the cotyledons and the epicotyl above the soil.

3. *The radicle is* found at the tip of the hypocotyl and produces the primary root when germination occurs.

4. *The general function of the cotyledon is to supply the embryo with* food. Some cotyledons store food. Others digest and absorb food stored in the endosperm. Some cotyledons emerge from the seed and manufacture food by photosynthesis.

5. *The coleorhiza protects the* radicle when it emerges from the seed. The coleorhiza is a sheath that surrounds the radicle and root tip of embryos in grass seeds. After the root tip has emerged from the seed the coleorhiza has no further function.

6. *The first part to emerge from a germinating seed is the* radicle, which develops into the primary root. This permits the seedling to become anchored in the soil before other parts emerge.

7. *Afterripening of a seed occurs when* the seed undergoes physiological changes. Many seeds will not germinate until these changes have been completed.

8. *In the strictest sense of the term a fruit is a ripened* ovary. Some fruits are derived from several ovaries, and in some species the receptacle or floral parts are so closely associated with the fruit as to become essentially part of it. In practice the term fruit is used to include these cases.

9. *A drupe is a* fleshy, one-seeded fruit with a hard endocarp. Examples are peaches and plums in which the hard, stony endocarp encloses the single seed.

10. *Dry fruits are* either dehiscent (splitting open at maturity) or indehiscent (not splitting open at maturity).

11. *A strawberry is an* aggregate-accessory fruit. It consists of many simple fruits on an enlarged, ripened receptacle.

12. *Wind-disseminated seeds and fruits tend to have* hairs or wings by which they can be easily lifted and blown by wind.

Answers

d	1
c	2
a	3
c	4
d	5
b	6
d	7
c	8
a	9
c	10
d	11
b	12

Development of Fruit Parts from Flower Parts

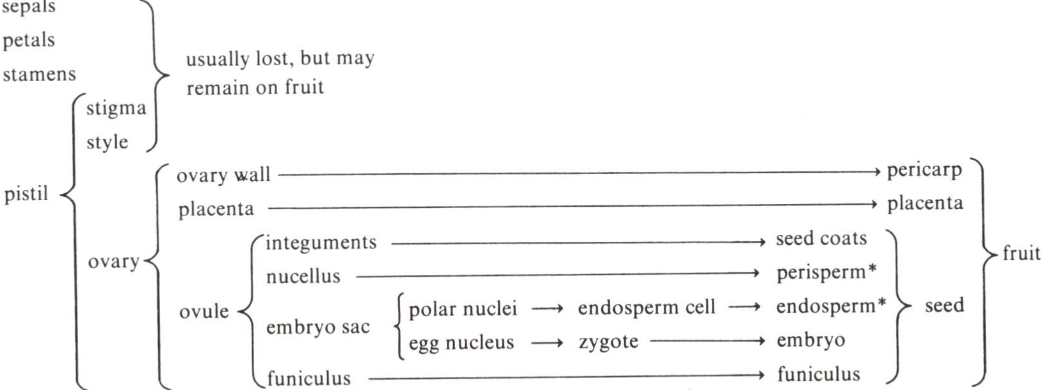

*May not be present.

GREGATE FRUIT develops from the many simple pistils of a single flower. A MULTIPLE FRUIT is derived from the many pistils of many closely associated flowers. ACCESSORY FRUITS are those that include the floral tube or stem tissue; they may be SIMPLE-ACCESSORY (apple, which includes the floral tube), AGGREGATE-ACCESSORY (strawberry, which includes the receptacle on which the flower grew), or MULTIPLE-ACCESSORY (pineapple, which includes the floral tubes of the individual flowers and the peduncle on which they grew).

Fruits and seeds may be DISPERSED by wind, animals, or water, or they may be self-dispersed. Wind-disseminated seeds and fruits are light and are often equipped with wings or hairs. Those disseminated by animals may have hooks, barbs, or sticky seed coats by means of which they become attached to the hair or feet of animals; they are carried some distance before they drop off or are shed with the fur. Other animal-disseminated fruits and seeds are eaten and pass through the digestive system. Some fruits open and shoot their seeds several feet away from the parent plant.

When they are shed, many seeds go into a period of DORMANCY, during which they have low water content and a slow rate of metabolism. Dormancy can be broken only by certain conditions. In some cases exposure to light may be necessary for germination; in others it may be alternate freezing and thawing or abrasion (SCARIFICATION) of a thick seed coat. Nearly all seeds require oxygen and a supply of water for germination. Some may require AFTERRIPENING, a series of internal physiological changes.

Simple fruits may be fleshy or dry. Simple FLESHY FRUITS are those which have moist, often juicy, pericarps; simple DRY FRUITS have woody or papery pericarps. Simple fleshy fruits are either drupes, berries, or pomes. A DRUPE (peach, olive) comes from a simple ovary with one ovule; the exocarp and mesocarp are fleshy, but the endocarp is stony, forming a pit. A BERRY (tomato, grape) differs from a drupe in being derived from a compound ovary, in containing numerous seeds, and in having a completely fleshy pericarp. Some special types of berries are the HESPERIDIUM (orange, lemon) which has a leathery rind and the PEPO (squash, watermelon) which has a hard rind. A POME (apple, pear) has an edible portion composed largely of floral parts that surround the ovary. The core is formed from the endocarp.

Dry fruits are either DEHISCENT (splitting open at maturity and usually many-seeded) or INDEHISCENT (not splitting open at maturity and usually one-seeded). Some dehiscent dry fruits are capsules, legumes, and follicles. A CAPSULE is derived from a compound ovary and opens by lengthwise slits (lily, rhododendron), by transverse slits (plantain), or by pores (poppy). LEGUMES (bean, pea) and FOLLICLES (milkweed, larkspur) each are derived from a simple ovary, the former splitting open on two opposite sides, the latter on only one side. Indehiscent dry fruits are grains, achenes, samaras, nuts, and schizocarps. The seed in a GRAIN (corn, wheat), or CARYOPSIS, has a seed coat fused at all points to the pericarp. In an ACHENE (sunflower, dandelion), the seed is attached to the pericarp only by its funiculus. A SAMARA (maple, ash), or KEY, resembles an achene, but the pericarp bears a thin wing. A NUT (chestnut, oak) is also similar to an achene, but its pericarp is hard. A SCHIZOCARP (carrot) arises from a compound ovary, the carpels splitting apart at maturity; each segment is one-seeded and resembles an achene.

Seeds and fruits are rich in STORED FOODS. Starch is the most common carbohydrate of seeds. Sugars are often found in the pericarp of fruits, occasionally in seeds. Protein-rich seeds are common in the legume family. The endosperm of grass seeds has an outer layer, the ALEURONE, which is rich in protein. Fats and oils are found in many seeds and fruits. These foods can be used not only by germinating seedlings, but also by animals.

13 MEIOSIS AND NUCLEIC ACIDS

SELF-TEST

1. When a cell with 20 chromosomes undergoes meiosis, each of the four resulting cells has
   a 10 chromosomes
   b 20 chromosomes
   c 40 chromosomes
   d 5 chromosomes

2. Meiosis differs from mitosis in that in meiosis
   a chromosomes do not pair and their number is reduced
   b chromosomes do not pair and their number is unchanged
   c chromosomes pair and their number is reduced
   d chromosomes pair and their number remains unchanged

3. In meiosis homologous chromosomes separate from each other in
   a metaphase I
   b anaphase I
   c metaphase II
   d anaphase II

4. The genetic material of a cell is
   a protein
   b DNA
   c messenger RNA
   d transfer RNA

5. A copy of the genetic code is carried from the nucleus to the cytoplasm by
   a protein
   b DNA
   c messenger RNA
   d transfer RNA

6. The genetic code for a given amino acid consists of
   a one nucleotide
   b two nucleotides
   c three nucleotides
   d four nucleotides

7. DNA differs from RNA in being
   a double-stranded and containing deoxyribose and thymine
   b double-stranded and containing ribose and uracil
   c single-stranded and containing deoxyribose and thymine
   d single-stranded and containing ribose and uracil

8. DNA serves as a template for the synthesis of
   a protein
   b DNA
   c RNA
   d DNA and RNA

9. Replication of DNA is associated with
   a mitosis but not meiosis
   b meiosis but not mitosis
   c mitosis and the first division of meiosis
   d mitosis and the second division of meiosis

10. A mutation is a change in
    a DNA
    b RNA
    c chromosome structure
    d chromosome number

1 ________
2 ________
3 ________
4 ________
5 ________
6 ________
7 ________
8 ________
9 ________
10 ________

BASIC FACTS

Most of the cells of a seed plant are DIPLOID, that is, the chromosomes are present in pairs, one from each parent. The two members of a pair are called HOMOLOGOUS chromosomes. New diploid cells are produced by MITOSIS in which all the chromosomes are duplicated and identical copies go to the two daughter cells. The number of chromosomes in a diploid cell is usually constant for a species.

Gametes are HAPLOID and are produced by meiosis. MEIOSIS is a set of two nuclear divisions in which the chromosomes are duplicated once and in which four daughter cells are produced. The number of chromosomes per cell is reduced by half. The four daughter cells are not genetically identical with each other. Each one carries one chromosome of each homologous pair.

CHROMOSOMES consist primarily of protein and nucleic acid. Two types of NUCLEIC ACID are DEOXYRIBONUCLEIC ACID (DNA) and RIBONUCLEIC ACID (RNA). DNA is the genetic material and contains the genetic code that controls the production of enzymes and therefore the metabolism of the cell. A GENE is the portion of a DNA molecule that carries the code for one enzyme. Usually one DNA molecule carries the code for several enzymes. Genes occur in linear order on chromosomes.

Three types of RNA occur in cells: MESSENGER RNA (mRNA), TRANSFER RNA (tRNA), and RIBOSOMAL RNA. Coded information stored in DNA is transferred from the nucleus to the ribosomes in the cytoplasm by mRNA.

(Continued on page 52)

ADDITIONAL INFORMATION

Each of the two nuclear divisions of MEIOSIS is divided into four phases: prophase, metaphase, anaphase, and telophase. A Roman numeral I or II is appended to the phase to indicate the first or second division. In PROPHASE I the chromosomes are duplicated and each consists of two chromatids as in prophase of mitosis. But in prophase I, homologous chromosomes pair with each other, forming a TETRAD of chromatids. This pairing is called SYNAPSIS; it does not occur in mitosis. In METAPHASE I the pairs line up across the spindle, and in ANAPHASE I the homologous chromosomes separate from each other and move to opposite poles of the cell. In TELOPHASE I a nuclear membrane forms around each group of chromosomes and usually a cell wall separates the cell into two daughter cells. These daughter cells are not genetically identical. The separation of homologous chromosomes results in the segregation of genes that are on those chromosomes.

No duplication of chromosomes occurs between the two divisions. In PROPHASE II, each cell forms its own spindle. At METAPHASE II the chromosomes, each still consisting of two chromatids, line up on the spindle, and during ANAPHASE II the chromatids of each chromosome separate and move to opposite poles of the cells. Nuclear membranes form in TELOPHASE II, and cell walls divide the two cells into four daughter cells. Each daughter cell has half of the chromosomes that the parent cell had and possesses one member of each pair of homologous chromosomes.

DNA is a double-stranded macromolecule. Each strand consists of thousands of nucleotides each composed of a DEOXYRIBOSE (a 5-carbon sugar), a phosphate, and a nitrogen base that is either a purine (ADENINE or GUANINE) or a pyrimidine (CYTOSINE or THYMINE). The guanine of one strand is opposite the cytosine of the other strand and is loosely held to it by hydrogen bonds (weak chemical bonds). The adenine of one strand is similarly held to the thymine of the other. The two strands form a ladderlike molecule which can be illustrated by using the initials of the components:

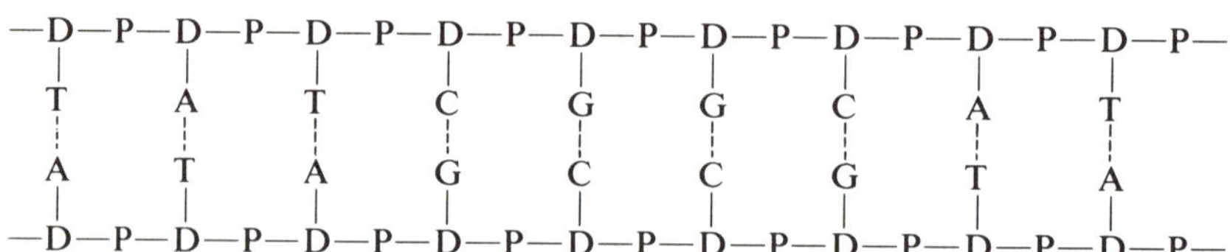

These strands are twisted around each other in the form of a double helix. When a DNA molecule replicates itself the hydrogen bonds break and the two strands separate from each other. Each strand serves as a template on which the missing half is synthesized. Thus two identical DNA molecules are produced.

(Continued on page 52)

EXPLANATIONS

1. *When a cell with 20 chromosomes undergoes meiosis, each of the four resulting cells has* 10 chromosomes. Meiosis results in cells which have half the number of chromosomes as the parent cell.

2. *Meiosis differs from mitosis in that in meiosis* chromosomes pair and their number is reduced. The chromosome number is reduced by half in meiosis because two nuclear divisions take place with only one duplication of chromosomes. The chromosomes are then distributed among the four resulting cells.

3. *In meiosis homologous chromosomes separate from each other in* anaphase I, and identical chromatids separate from each other in anaphase II.

4. *The genetic material of a cell is* DNA. DNA contains the genetic code that codes for the production of proteins.

5. *A copy of the genetic code is carried from the nucleus to the cytoplasm by* messenger RNA. It is on this RNA that amino acids assemble before being combined into protein molecules.

6. *The genetic code for a given amino acid consists of* three nucleotides. This triplet code contains 64 triplets, more than enough to code for the 20 naturally occuring amino acids.

7. *DNA differs from RNA in being* double-stranded and containing deoxyribose and thymine. RNA is single-stranded and contains ribose and uracil in place of deoxyribose and thymine. Both types of nucleic acid contain adenine, guanine, cytosine, and phosphate.

8. *DNA serves as a template for the synthesis of* both DNA and RNA. Both strands of DNA act as templates in the synthesis of additional DNA. One strand of DNA is the template in RNA synthesis.

9. *Replication of DNA is associated with* mitosis and the first division of meiosis. No increase in amount of genetic material is associated with the second division of meiosis.

10. *A mutation is a change in* DNA and is thereafter inherited like any other genetic material. Many, but not all, mutations are recessive and may be carried undetected for many generations even if lethal. The term mutation has been used in the past to include chromosomal changes, but it is not usually used so now.

Answers

Answer	No.
a	1
c	2
b	3
b	4
c	5
c	6
a	7
d	8
c	9
a	10

Much of the ribosome is composed of ribosomal RNA on which mRNA is positioned when it is functioning. The amino acids from which the enzyme is to be synthesized are brought to the ribosome by tRNA and are assembled in proper order on the mRNA molecule.

Preceding mitosis and the first division of meiosis (but not the second division of meiosis) each gene REPLICATES itself. One copy is present in each of the newly formed chromatids. In mitosis each of the two daughter cells receives one of these identical copies. In meiosis only two of the four daughter cells receive a copy of any one gene.

Ordinarily the replication of genes and chromosomes provides perfect copies of the originals, but occasionally a change or error occurs. A change in a gene is called a MUTATION. Changes in whole chromosomes include DELETION (the loss of a portion of a chromosome), INVERSION (the breaking off of part of a chromosome and its insertion back in the same chromosome so that the genes are in reverse order), DUPLICATION of a portion of a chromosome, and TRANSLOCATION (the exchange of broken fragments between nonhomologous chromosomes). Still another error is a change in chromosome number. POLYPLOIDY is the condition in which cells possess some multiple of the haploid number: triploid cells have the equivalent of three haploid sets of chromosomes, tetraploids have four haploid sets, pentaploids have five, etc. Such changes provide variety in the genetic material.

The GENETIC CODE is the order in which the nucleotides occur in one of the strands of the macromolecule. The code is called a TRIPLET CODE, for three adjacent nitrogen bases code for one amino acid. Each triplet of nitrogen bases is called a CODON. The order in which the codons occur in DNA determines the order in which amino acids are assembled in a particular protein.

DNA is confined almost entirely to the nucleus, and protein synthesis occurs in the cytoplasm. The coded information stored in DNA is transferred from the nucleus to the cytoplasm by messenger RNA (mRNA), a single-stranded nucleic acid. A strand of mRNA consists of phosphate and RIBOSE chemically bound in alternate positions. One nitrogen base (ADENINE, GUANINE, CYTOSINE, or URACIL) is attached to each ribose. A mRNA molecule is synthesized on one strand of DNA, which serves as a template. The order of nitrogen bases in mRNA reflects the order in the template strand of DNA: guanine and cytosine occur opposite each other, adenine occurs opposite the thymine of DNA, and uracil occurs opposite the adenine of DNA:

```
—D—P—D—P—D—P—D—P—D—P—D—P—D—P—D—P—D—P—
 |    |    |    |    |    |    |    |    |     one strand of DNA
 T    A    T    C    G    G    C    A    T

 A    U    A    G    C    C    G    U    A
 |    |    |    |    |    |    |    |    |     mRNA
—R—P—R—P—R—P—R—P—R—P—R—P—R—P—R—P—R—P—
```

A molecule of mRNA thus carries a copy of the code. The mRNA molecules move to the ribosomes in the cytoplasm. Amino acids are assembled into position according to the order in which the codons occur on the mRNA molecule. For example, the codon AUA codes for one amino acid, GCC for another, and GUA for another. With four nitrogen bases, sixty-four different codons are possible. Since only twenty naturally occuring amino acids need be coded for, several codons code for the same amino acid. Condensation reactions unite the amino acids into a protein molecule which has enzymatic properties.

A MUTATION is a change in the DNA molecule. More specifically it is the addition, deletion, or substitution of one nitrogen base. This changes one codon into another which codes for a different amino acid. A change of one amino acid in a protein changes it to another protein. The new protein may be completely inactive enzymatically, or occasionally it may have different enzymatic properties. In either case it can effect a change in the metabolism of the cell. A mutant strand of DNA is replicated and inherited as is normal DNA.

14 HEREDITY

SELF-TEST

1. A sexually reproducing organism contributes to each of its offspring
   a $\frac{1}{4}$ of its genes
   b $\frac{1}{2}$ of its genes
   c $\frac{3}{4}$ of its genes
   d all of its genes

2. Mendel's second law is known as the law of
   a segregation
   b independent assortment
   c recombination
   d incomplete dominance

3. Alleles are
   a homologous chromosomes
   b chromosomes that have crossed over
   c linked genes
   d alternate forms of a gene

4. Independent assortment occurs between genes
   a in two different individuals
   b in two different cells
   c on two different pairs of chromosomes
   d on the same chromosome

5. A plant heterozygous for two pairs of genes on two different pairs of chromosomes can produce
   a two kinds of gametes
   b four kinds of gametes
   c six kinds of gametes
   d eight kinds of gametes

6. The inheritance of a trait from the female parent only and never from the male is probably the result of
   a monohybrid crossing
   b incomplete dominance
   c multiple gene inheritance
   d cytoplasmic inheritance

For the following questions you are given that tall (T) is dominant to dwarf (t) and axial flowers (A) are dominant to terminal flowers (a).

7. In the cross TT × tt, the F_2 generation is expected to have a phenotype ratio of
   a 3 tall:1 dwarf
   b 3 dwarf:1 tall
   c 1 tall:1 dwarf
   d all tall

8. In the cross TTAA × ttaa, the F_2 generation is expected to have a phenotype ratio of
   a 9 dwarf axial:3 dwarf terminal: 3 tall axial:1 tall terminal
   b 9 dwarf terminal:3 tall axial: 3 tall terminal:1 dwarf axial
   c 9 tall axial:3 tall terminal: 3 dwarf axial:1 dwarf terminal
   d 9 tall terminal:3 dwarf axial: 3 dwarf terminal:1 tall axial

9. If a dwarf plant with terminal flowers produces 50 tall offspring with axial flowers and 50 tall offspring with terminal flowers, the genotype of the other parent plant is
   a TTAA
   b TTAa
   c TtAA
   d TtAa

1 ______
2 ______
3 ______
4 ______
5 ______
6 ______
7 ______
8 ______
9 ______

BASIC FACTS

The study of heredity is called GENETICS. The basic laws of heredity were discovered by GREGOR MENDEL in the nineteenth century.

Like chromosomes, genes occur in pairs. Both genes of a pair affect the same character, but they are not necessarily identical. For instance, in peas there is a gene for yellow cotyledons and a gene for green cotyledons. An individual with identical genes for one character (such as two genes for yellow or two for green) is HOMOZYGOUS; one with nonidentical genes for a character (one for yellow and one for green) is HETEROZYGOUS (HYBRID). Two or more alternate forms of a gene are ALLELES.

If in a heterozygous individual only one gene finds expression, that gene is DOMINANT, and the other is RECESSIVE. Commonly the capitalized initial of the dominant trait is used to indicate the dominant gene, and the same initial in lower case indicates the recessive gene (e.g., Y for yellow, y for green). The total genetic constitution of an individual is its GENOTYPE. The PHENOTYPE is the visible manifestation of a genotype. Thus an embryo with the genotype YY or Yy is phenotypically yellow. An embryo with the phenotype green has the genotype yy.

In a cross of two homozygous individuals, YY and yy, the former produces only gametes with Y and the latter produces only gametes with y; all their offspring (called F_1) have the genotype Yy. Half of the eggs and half of the sperms of the F_1 generation carry Y and half of the eggs and half of the sperms carry y. If an F_1 plant is self-pollinated, four different unions of gametes are possible and all four are equally probable: both sperm and egg have gene Y, egg has Y and sperm has y, egg has y and sperm has Y, or both egg and sperm have y. These unions produce three different genotypes in the F_2 (second generation

(Continued on page 56)

ADDITIONAL INFORMATION

Since genes are located on chromosomes they have distribution patterns in MEIOSIS similar to those of chromosomes. After meiosis each of the four resulting haploid cells contains one gene of each pair. In a heterozygous individual half of the gametes have the dominant gene and half have the recessive gene. This separation of genes in meiosis is called SEGREGATION (MENDEL'S FIRST LAW).

When paired chromosomes arrange themselves across the equator in metaphase I, it is chance that determines which chromosome of the pair is oriented toward either pole of the cell. The segregation of one pair of chromosomes (and their genes) does not influence the segregation of another pair. Thus in an individual heterozygous for two pairs of genes on two pairs of chromosomes, both dominant genes may segregate from both recessive genes at anaphase I, or one dominant and one recessive gene may segregate from their corresponding recessive and dominant genes. In such an individual $\frac{1}{4}$ of the gametes carry both dominant genes, $\frac{1}{4}$ both recessives, $\frac{1}{4}$ one dominant and one recessive, and $\frac{1}{4}$ the other dominant and the other recessive. This is called INDEPENDENT ASSORTMENT (MENDEL'S SECOND LAW).

Independent assortment does not occur between pairs of genes that are on the same pair of chromosomes. Such genes are LINKED and are inherited together unless crossing over occurs between them. CROSSING OVER is the exchange of similar parts between chromatids of homologous chromosomes when they pair in prophase I of meiosis. Since crossing over is more frequent between genes that are farther apart and less frequent between those that are close, the positions of genes on chromosomes can be calculated by determining the frequency of crossing over between genes. From this information CHROMOSOME MAPS are made.

In cases of INCOMPLETE DOMINANCE (sometimes called BLENDING), one gene of a pair is not dominant to the other, and heterozygous individuals show an intermediate phenotype. For example, snapdragons heterozygous for red and white flower color have pink flowers. Both red- and white-flowered plants are homozygous, pink-flowered plants are heterozygous. The offspring of a pink × pink cross are of the phenotype ratio $\frac{1}{4}$ red, $\frac{1}{2}$ pink, and $\frac{1}{4}$ white.

In cases of MULTIPLE GENE INHERITANCE, several pairs of genes influence a single character. The ratios obtained in these crosses are fairly complex and depend in part on the number of pairs of genes involved. Often the differences between phenotypes are so slight that they seem to represent a continuum.

Some genetic material is found in the cytoplasm. Both chloroplasts and mitochondria are known to have a small

(Continued on page 56)

EXPLANATIONS

1. *A sexually reproducing organism contributes to each of its offspring* half of its genes. Each egg or sperm contains half of the chromosomes of the organism and therefore half of its genes.

2. *Mendel's second law is known as the law of* independent assortment. It states that the segregation of genes on one pair of chromosomes is independent of the segregation of genes on other pairs.

3. *Alleles are* alternate forms of a gene. Alleles occupy the same positions on homologous chromosomes.

4. *Independent assortment occurs between genes* on two different pairs of chromosomes in the same organism.

5. *A plant heterozygous for two pairs of genes on two different pairs of chromosomes can produce* four kinds of gametes. A plant heterozygous for one pair of genes can produce two kinds of gametes, half the gametes carrying one gene and half carrying the other. Each additional pair of genes on a different pair of chromosomes doubles the number of types of gametes the organism can produce.

6. *The inheritance of a trait from the female parent only and never from the male is probably the result of* cytoplasmic inheritance since cytoplasm is inherited from the female parent only.

7. *In the cross TT × tt, the F_2 generation is expected to have a phenotype ratio of* 3 tall:1 dwarf. In a cross between an individual homozygous for a dominant gene and an individual homozygous for a recessive gene, $\frac{3}{4}$ of the F_2 generation have the dominant phenotype and $\frac{1}{4}$ have the recessive phenotype.

8. *In the cross TTAA × ttaa, the F_2 generation is expected to have a phenotype ratio of* 9 tall axial: 3 tall terminal: 3 dwarf axial: 1 dwarf terminal. In a cross between an individual homozygous for two dominant genes and an individual homozygous for two recessive genes, $\frac{9}{16}$ of the F_2 generation have both dominant phenotypes, $\frac{3}{16}$ have one dominant and one recessive phenotype, $\frac{3}{16}$ have the other dominant and the other recessive phenotype, and $\frac{1}{16}$ have both recessive phenotypes.

9. *If a dwarf plant with terminal flowers produces 50 tall offspring with axial flowers and 50 tall offspring with terminal flowers, the genotype of the other parent plant is* TTAa. Since none of the offspring is dwarf it is unlikely that the other parent has the gene t. Half of the offspring are homozygous for terminal flowers; therefore they must have received the gene a from the other parent. Half of the offspring have axial flowers and therefore possess the gene A which could have been received only from the other parent.

Answers

b	1
b	2
d	3
c	4
b	5
d	6
a	7
c	8
b	9

Some Ratios Associated with Monohybrid Crosses

Cross	*Phenotype Ratio, Complete Dominance*	*Genotype Ratio*
AA × AA	all dom.	all AA
AA × Aa	all dom.	1 AA : 1 Aa
AA × aa	all dom.	all Aa
Aa × Aa	3 dom. : 1 rec.	1 AA : 2 Aa : 1 aa
Aa × aa	1 dom. : 1 rec.	1 Aa : 1 aa
aa × aa	all rec.	all aa

of offspring): $\frac{1}{4}$ have genotype YY, $\frac{1}{2}$ have genotype Yy, and $\frac{1}{4}$ have genotype yy (genotype ratio 1:2:1). Those with genotypes YY and Yy comprise $\frac{3}{4}$ of the offspring and are phenotypically yellow; the remaining $\frac{1}{4}$ are green. The phenotype ratio of three dominant to one recessive (3:1) is typical of the F_2 of a MONOHYBRID CROSS (a cross involving one pair of genes). A monohybrid cross illustrates MENDEL'S FIRST LAW (law of segregation) which may be stated: the two alleles of a pair do not blend or contaminate each other but segregate from each other during meiosis and pass into different gametes.

In a cross between two individuals of genotypes YYRR (yellow, rounded cotyledons) and yyrr (green, wrinkled cotyledons), the gametes produced by the former individual are all YR, and those produced by the latter are all yr. Their F_1 offspring are all yellow and round, for they are heterozygous for both pairs of genes, YyRr. Each gamete of an F_1 individual carries one gene from each pair of genes. Four kinds of eggs and four kinds of sperms are possible and are produced in equal proportions: $\frac{1}{4}$ YR, $\frac{1}{4}$ Yr, $\frac{1}{4}$ yR, and $\frac{1}{4}$ yr. When fertilization occurs any combination of egg and sperm is possible, and all combinations are equally probable. The expected phenotype ratio of the F_2 is $\frac{9}{16}$ yellow round, $\frac{3}{16}$ yellow wrinkled, $\frac{3}{16}$ green round, and $\frac{1}{16}$ green wrinkled. A 9:3:3:1 phenotype ratio is typical of the F_2 of a DIHYBRID CROSS (a cross involving two pairs of genes). A dihybrid cross illustrates MENDEL'S SECOND LAW (law of independent assortment) which may be stated: at meiosis pairs of genes on different pairs of chromosomes segregate independently of each other.

amount of DNA, and some chloroplast characteristics are inherited through the cytoplasm. Since the male parent usually contributes only the sperm nucleus to the offspring, CYTOPLASMIC INHERITANCE shows a pattern in which traits are passed from the female parent to the offspring, but never from the male parent to the offspring.

Only part of the genotype of a plant showing the dominant phenotype is immediately known. A yellow-seeded plant carries one gene Y, but the other gene may be either Y or y. To determine the identity of the second gene, a TEST CROSS (BACK CROSS) is performed. The plant is crossed with one possessing the recessive phenotype (green in this case) and therefore known to be homozygous (yy). If half of the offspring are yellow and half are green, the original plant is heterozygous, for the green offspring must have obtained a recessive gene, y, from each parent. If all the offspring are yellow, it is concluded that the original plant is homozygous.

INBREEDING leads to increased homozygosity; highly inbred individuals often are weak and have stunted growth. OUTBREEDING leads to heterozygosity and to an increase in vigor. HYBRID VIGOR is shown by offspring of crosses between different varieties or species. Increased yields in corn harvests have been obtained from HYBRID CORN. Two lines of inbred corn are crossed to produce F_1 hybrids; these hybrids are crossed with F_1 hybrids obtained from two other inbred lines. The seeds resulting from this "double cross" are planted by farmers.

In general, the environment does not affect the genotype of an individual, but it may affect the phenotype; for example, plants genetically capable of forming chlorophyll are colorless if raised in the dark.

To work out problems involving more than one pair of genes one need merely consider each pair separately, determine the ratio expected from each pair of genes, and then "multiply" these ratios. Thus the problem to determine the phenotypic ratio from the cross YyRr × YyRr becomes

  3 yellow:1 green
× 3 round: 1 wrinkled
9 yellow round:3 yellow wrinkled:3 green round:1 green wrinkled

If three pairs of genes are involved, work out the ratio for any two pairs, then multiply the answer by the ratio for the third pair. This method gives the same answers as the checkerboard method so often used, and it is somewhat shorter. It can also be used to determine genotypic ratios.

15 GROWTH, DIFFERENTIATION, AND TROPISMS

SELF-TEST

1. A substance produced in one part of an organism and transported to another part where it exerts an effect is a(n)
   a hormone
   b porphyrin
   c enzyme
   d cytoplasmic gene

2. Stem elongation is known to be affected by
   a auxins and gibberellin
   b gibberellin and florigen
   c florigen and kinins
   d kinins and auxins

3. A tropism is a movement
   a exhibited only by motile cells
   b caused by a sudden change in the turgor pressure of cells
   c caused by internal stimuli
   d caused by unequal growth on opposite sides of an organ

4. The opening and closing of flowers are
   a autonomic movements
   b nastic movements
   c nutations
   d tropisms

5. The twining of a stem may involve
   a thigmotropism
   b nutation
   c differential growth
   d all of the above

6. With respect to tropisms stems are
   a positively phototropic and positively geotropic
   b positively phototropic and negatively geotropic
   c negatively phototropic and positively geotropic
   d negatively phototropic and negatively geotropic

7. The hormone that influences tropisms is
   a florigen
   b kinetin
   c gibberellin
   d IAA

8. The flowering of carrots in July is due to
   a gradual increase of temperature during the preceding months
   b long days and short nights
   c alternate dry and rainy weather

9. The light receptor in photoperiodism is
   a cytochrome
   b phytochrome
   c chlorophyll b
   d carotene

10. Genetic material is distributed among cells in such a way that each cell has
    a only those genes that are necessary for the development of the tissue of which it is a part
    b a complete set of genes, all of them being active during the life of the cell
    c a complete set of genes, only a few of them being active at a time

1 ______
2 ______
3 ______
4 ______
5 ______
6 ______
7 ______
8 ______
9 ______
10 ______

BASIC FACTS

GROWTH is an increase in the number and size of cells in an organism; it usually includes their DIFFERENTIATION into specialized cells having specific functions. Growth is determined by both environmental and hereditary factors; it is regulated internally by hormones. MORPHOGENESIS includes all of the metabolic and anatomical changes that produce the various organs of an individual.

The ENVIRONMENTAL FACTORS that affect plant growth include light, temperature, water content of the air and soil, and the availability of minerals.

LIGHT affects the rate of growth and the color of stems and leaves. Plants grown in constant darkness have long, weak stems; their leaves fail to expand; and they are either white or pale yellow. Plants grown in bright light have shorter, sturdier stems and expanded leaves; they are usually green.

The flowering of many plants is determined by day and night lengths. This response is called PHOTOPERIODISM. Plants that bloom only under short days and long nights are called SHORT-DAY PLANTS; those that bloom only under long days and short nights are called LONG-DAY PLANTS. DAY-NEUTRAL PLANTS bloom no matter what the day length.

The optimum TEMPERATURE for growth usually lies between 10° and 40°C, growth being slower at higher or lower temperatures. In some species flowering is either modified or controlled by temperature. Exposure to low temperature is required for flowering in biennial plants and for germination of seeds or opening of buds in some species.

HORMONES are substances that are produced in one part of the plant and that move to another part where, through their effects on metabolism, they control growth, differentiation, and tropisms. The three main types of plant hormones are auxins, gibberellins, and kinins. AUXINS play several

(Continued on page 60)

ADDITIONAL INFORMATION

A zygote and all the diploid cells that develop from it contain a full complement of identical genetic material. The DIFFERENTIATION of embryonic cells into tissues and organs with diverse properties is due to the activity of only certain genes within the different cells. The formation of a mature plant with all of its organs depends on the proper ordering of chemical reactions. HORMONES accomplish this integration by controlling the activity of specific genes in different cells. VITAMINS have enzymatic functions, and some of them have hormonal properties.

The receptive organ in PHOTOPERIODISM is the leaf. Only the leaves need be exposed to the proper day length for a plant to bloom. It is inferred that under the proper day length FLORIGEN is synthesized in the leaf and transported to the stem tip where it converts a vegetative apical meristem into a reproductive one. The photoreceptive pigment in the leaf is PHYTOCHROME, which is a blue-green pigment existing in two forms. One form, designated P_{660}, absorbs red light of wavelength 660 mμ and is immediately converted to the other form, P_{730}. P_{730} absorbs light of wavelength 730 mμ (called far red); it is immediately converted to P_{660}. In darkness P_{730} slowly reverts to P_{660}. During daylight hours phytochrome is in the form of P_{730} because of the high red light content of sunlight. At night P_{730} is converted slowly to P_{660}. How nearly complete the conversion is depends in part on the length of the night. It is believed that the state of phytochrome (whether P_{730} or P_{660}) during the night hours determines whether or not florigen is produced.

The effect of TEMPERATURE on plants lies in its effect on the individual processes occurring within the plant. Each process has its own minimum, optimum, and maximum temperatures. The amount and type of growth of a particular plant depend on the effect of temperature on all its chemical and physical processes.

One of the best known AUXINS is INDOLE-3-ACETIC ACID (IAA), a growth hormone that is produced by active meristems, particularly young leaves and apical meristems. IAA moves downward from the stem tip and causes enlargement of cells in the region of elongation. IAA produced in the root tip moves upward and exerts a similar effect on root cells in the region of elongation. Other functions of IAA include the initiation of cambial activity in the spring and the prevention of premature abscission of leaves and fruits. The early enlargement of an ovary as it develops into a fruit is brought about by auxin introduced by pollen grains. Continued growth of fruits is due to auxin produced by the seeds developing within the fruit. Auxins together with kinins play a role in differentiation of organs and tissues. The relative concentrations of these two substances determine whether roots or buds are produced in a particular region of the plant.

(Continued on page 60)

EXPLANATIONS

1. *A substance produced in one part of an organism and transported to another part where it exerts an effect is a* hormone. Hormones control many metabolic reactions in cells. Hormones are probably all enzymatic, but not all enzymes are hormones.

2. *Stem elongation is known to be affected by* auxins and gibberellin, two naturally occurring plant hormones.

3. *A tropism is a movement* caused by unequal growth on opposite sides of an organ and is dependent on unilateral presentation of an external stimulus. It occurs only where cells are elongating; therefore bending is associated with the region of elongation. The responding organ characteristically grows toward or away from the source of the stimulus.

4. *The opening and closing of flowers are* nastic movements. Nastic movements of petals, sepals, and leaves are due to differences in turgor pressure or in growth rates on the upper and lower surfaces of the organ.

5. *The twining of a stem may involve* thigmotropism or nutation, both of which are caused by differential growth. Thigmotropism is a response to touch, an external stimulus; a nutation is a spontaneous movement and requires no external stimulus.

6. *With respect to tropisms stems are* positively phototropic and negatively geotropic. Roots are positively geotropic; some are negatively phototropic, and others are not responsive to light.

7. *The hormone that influences tropisms is* IAA. IAA causes increase in length of organs; unequal distribution causes unequal growth and thus bending of the affected organs.

8. *The flowering of carrots in July is due to* long days and short nights. Many plants bloom in response to certain critical lengths of day and night, and they characteristically bloom only during certain months of the year.

9. *The light receptor in photoperiodism is* phytochrome. This pigment is blue-green, but it is present in such low concentrations that it gives essentially no color to the cells possessing it.

10. *Genetic material is distributed among cells in such a way that each cell has* a complete set of genes, only a few of them being active at a time. Hormones control and coordinate the activity of genes.

Answers

a	1
a	2
d	3
b	4
d	5
b	6
d	7
b	8
b	9
c	10

roles in plants including the elongation of cells and the inhibition of growth of lateral buds. GIBBERELLINS cause elongation of stems and promote flowering in some species. KININS stimulate cell division. A flowering hormone called FLORIGEN is believed to exist, but it has not yet been isolated. The production of various hormones is controlled by the interaction of environmental conditions and the heredity of the plant.

GROWTH MOVEMENTS are caused by unequal growth on opposite sides of organs. Plants exhibit two types of growth movements: INDUCED, or PARATONIC, MOVEMENTS (tropisms and nastic movements), which are influenced by external stimuli, and SPONTANEOUS, or AUTONOMIC, MOVEMENTS (such as nutations), which are controlled by internal stimuli.

TROPISMS are responses to any of several stimuli including light (PHOTOTROPISM), gravity (GEOTROPISM), touch (THIGMOTROPISM), and chemicals (CHEMOTROPISM) that come primarily from one direction. Stems are positively phototropic and usually are negatively geotropic; they bend toward a bright light source, but most of them bend away from the earth. Roots are either negatively phototropic or do not respond to light; they are positively geotropic. Tendrils of many climbing plants are positively thigmotropic; they continue to grow around an object with which they have made contact. The growth of a pollen tube toward an ovule is a positive chemotropism.

NASTIC MOVEMENTS (such as the daily rising and lowering of leaves and petals) are responses to changes in light or temperature, but they are independent of the direction from which the stimulus comes.

NUTATION is the spiral movement made by a stem tip as it grows in space.

As long as the terminal bud remains on a stem, the lateral buds do not open or produce branches. This inhibition of lateral buds by terminal buds is called APICAL DOMINANCE. The IAA that is produced by the terminal bud and that moves down the stem is the inhibiting substance. If the terminal bud is removed or damaged, the uppermost of the lateral buds functions as a terminal bud. It produces its own IAA which causes elongation of the branch of which it is a part and inhibits the growth of lower lateral buds.

Application of GIBBERELLINS to some dwarf varieties of plants produces normal or nearly normal elongation of stems. Biennials, such as carrots, normally produce flowering stalks during their second year, but they may be induced to flower during their first year by application of gibberellins.

SYNTHETIC AUXINS are used commercially to produce seedless fruits in some plants and to induce root formation in stem cuttings for asexual propagation. One synthetic auxin, 2,4-dichlorophenoxyacetic acid (2,4-D), is used as a weed killer in lawns and fields of cereals. At the proper concentration 2,4-D upsets the hormonal control of most dicots and causes their death but has little effect on most monocots including grasses.

TROPISMS are associated with unequal concentrations of IAA on opposite sides of the responding organ. When a stem is unilaterally illuminated, the shaded side develops a higher concentration of IAA than does the illuminated side. This causes more rapid elongation of cells on the shaded side, and the stem bends toward the light source. Like stems, petioles are positively phototropic. Their responses tend to place leaf blades so that few leaves overlap and all leaves are as evenly illuminated as possible. This type of arrangement is called a LEAF MOSAIC. Horizontally placed stems and roots accumulate IAA on their lower sides, but stems bend upward and roots bend downward. The IAA concentration in roots is at or near the optimum for cell elongation there. Any additional accumulation of IAA on the shaded side of a root is likely to cause a decrease in the growth rate. For this reason the shaded side of those roots that do respond to light grows slower than the illuminated side and the root bends away from the light source.

Plants exhibit POLARITY; one end of the axis of an organ is physiologically and sometimes anatomically different from the other end.

16 INTRODUCTION TO THE PLANT KINGDOM

SELF-TEST

1. Three taxa that are lower than order are
   a division, genus, variety
   b class, division, species
   c family, species, genus
   d variety, class, family

2. Two plants that are members of two different orders in the same class are considered to be more closely related than two plants that are members of two different
   a varieties in the same species
   b divisions in the same kingdom
   c genera in the same family
   d species in the same genus

3. The complete scientific name of a plant includes
   a names of all the taxa of which the plant is a member
   b family, genus, and species names
   c genus and species names
   d genus and species names and the name of the person who first described the plant

4. Two features always present in the life cycles of sexually reproducing plants are
   a a sporophyte and a gametophyte
   b a sporophyte and meiosis
   c a sporophyte and fertilization
   d meiosis and fertilization

5. A gametophyte typically is
   a haploid and produces gametes
   b haploid and is produced by gametes
   c diploid and produces gametes
   d diploid and is produced by gametes

6. Members of the Embryophyta have in their life cycles
   a gametophytes only
   b sporophytes only
   c both gametophytes and sporophytes
   d neither gametophytes nor sporophytes

7. The two subkingdoms of the plant kingdom are
   a Tracheophyta and Thallophyta
   b Thallophyta and Embryophyta
   c Embryophyta and Bryophyta
   d Bryophyta and Tracheophyta

8. A plant having single-celled reproductive organs, no vascular tissue, and no true roots, stems, or leaves would be classified under
   a Bryophyta
   b Embryophyta
   c Thallophyta
   d Tracheophyta

9. Seed plants include
   a club mosses and gymnosperms
   b gymnosperms and angiosperms
   c angiosperms and ferns
   d ferns and club mosses

10. In the most highly evolved plants
    a the sporophyte is the dominant generation
    b the gametophyte is the dominant generation
    c neither generation is dominant
    d there are three generations

1 ______
2 ______
3 ______
4 ______
5 ______
6 ______
7 ______
8 ______
9 ______
10 ______

BASIC FACTS

Taxonomists classify plants according to a system with seven major TAXA (singular, TAXON), or units of classification: species, genus, family, order, class, division (or phylum), and kingdom.

A SPECIES (plural, SPECIES) is a group of individuals that freely breed among themselves and produce fertile offspring. Crosses may occur between members of two closely related species, but usually the offspring are sterile. There is no limit to the number of individuals in a species or to the area the species may occupy.

Species considered to be closely related, that is, to have common ancestors, are grouped into a GENUS (plural, GENERA). Closely related genera are grouped into a FAMILY. Closely related families comprise an ORDER. Closely related orders form a CLASS. Closely related classes are grouped into a DIVISION. The division is the highest taxon in the plant kingdom and corresponds to the phylum in animal taxonomy. All divisions (or phyla) are grouped into one of two KINGDOMS: the plant kingdom and the animal kingdom. Since these two kingdoms are not sharply delineated, some taxonomists prefer to classify the simplest plants and animals in a third kingdom (Protista). The classification of any organism includes these seven taxa, but whenever it is convenient any of the seven major taxa may be divided into subgroups. SUBKINGDOMS, for instance, are subgroups of a kingdom; they rank lower than kingdom and higher than division. A SUBDIVISION ranks lower than division and higher than order, and so on. The term VARIETY is usually used for a subdivision of a species.

The SCIENTIFIC NAME of any organism consists of its genus and species. Because the name has two parts, it is called a BINOMIAL. All members of the same species have the same binomial. According to international rules of

(Continued on page 64)

ADDITIONAL INFORMATION

The simplest possible LIFE CYCLE is that of an ASEXUALLY REPRODUCING plant. A single-celled plant divides mitotically, forming two cells which become adult plants; a multicellular plant fragments into two or more parts. The blue-green algae are the only major group of plants reproducing asexually only.

SEXUAL LIFE CYCLES all involve the alternation of fertilization with meiosis, and all have at least one haploid stage (gametes) and at least one diploid stage (zygote). The simplest possible sexual life cycle is one in which a zygote undergoes meiosis, producing four haploid spores which are capable of acting immediately as gametes. These gametes fuse and form new zygotes. But in many plants fertilization is delayed and the haploid spores produce a multicellular, haploid plant by mitotic divisions. This haploid generation is called a GAMETOPHYTE because it produces gametes. Many algae have such a life cycle (Fig. 1). If a delay occurs before meiosis rather than before fertilization, a diploid generation results. This generation is called a SPOROPHYTE because it produces spores. A few plants have such a life cycle (Fig. 2).

ALTERNATION OF GENERATIONS occurs in those plants having both sporophyte and gametophyte generations (Fig. 3). Alternation of generations is characteristic of all higher plants and many algae and some fungi. Some algae have two generations which are identical in general appearance (ISOMORPHIC alternation of generations). In all members of the Embryophyta and in some members of the Thallophyta the two generations are different in size and general appearance (HETEROMORPHIC alternation of generations), and frequently one generation is an independent plant with the other generation dependent on it. In HOMOTHALLIC species the gametophyte is capable of self-fertilization; a single gametophyte can produce zygotes. In HETEROTHALLIC species separate gametophytes (either male and female or of plus and minus mating types) are produced, and both are required for sexual

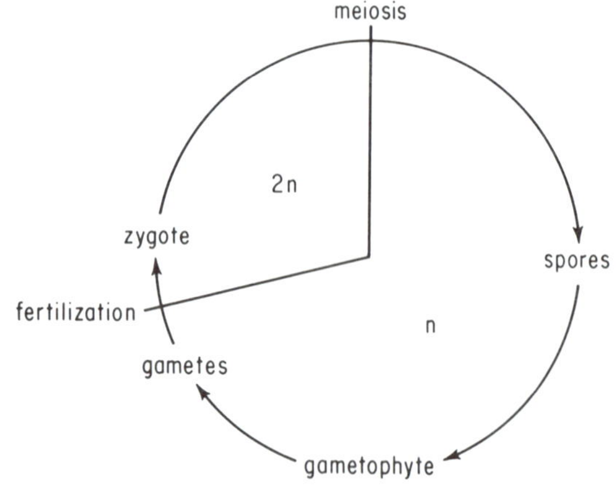

Fig. 1. Summary of a Life Cycle with Only a Gametophyte Generation

(Continued on page 64)

EXPLANATIONS

1. *Three taxa that are lower than order are* family, species, and genus. The seven main ranks of taxa, beginning with the highest, are: kingdom, division, class, order, family, genus, and species. Species are sometimes divided into varieties.

2. *Two plants that are members of two different orders in the same class are considered to be more closely related than two plants that are members of two different* divisions in the same kingdom. Our system of classification attempts to reflect the natural relationships among organisms. All species within a genus have common ancestors; all genera within a family have common ancestors, etc.

3. *The complete scientific name of a plant includes* genus and species names and the name of the person who first described the plant. In common usage only genus and species names are employed. Other taxa need not be used in naming a plant. Since a binomial is applied to only one species it is all that is needed to identify that species.

4. *Two features always present in the life cycles of sexually reproducing plants are* meiosis and fertilization. Every sexual cycle has haploid gametes and diploid zygotes but it does not necessarily have a gametophyte or a sporophyte generation.

5. *A gametophyte typically is* haploid and produces gametes. A gametophyte is produced by the germination of a haploid, sexual spore or by asexual reproduction of another gametophyte.

6. *Members of the Embryophyta have in their life cycles* both gametophytes and sporophytes. Alternation of these two generations is typical of all embryophytes. Some thallophytes have alternation of generations.

7. *The two subkingdoms of the plant kingdom are* Thallophyta (algae and fungi) and Embryophyta. The Bryophyta (mosses and liverworts) and the Tracheophyta (vascular plants) are the two divisions of the Embryophyta.

8. *A plant having single-celled reproductive organs, no vascular tissue, and no true roots, stems, or leaves would be classified under* Thallophyta. All members of the Embryophyta have multicellular reproductive organs; the Tracheophyta have vascular tissue and true roots, stems, and leaves.

9. *Seed plants include* gymnosperms and angiosperms. These are the only seed-bearing plants.

10. *In the most highly evolved plants* the sporophyte is the dominant generation. This is true of all Tracheophyta as well as of some algae.

Answers

c	1
b	2
d	3
d	4
a	5
c	6
b	7
c	8
b	9
a	10

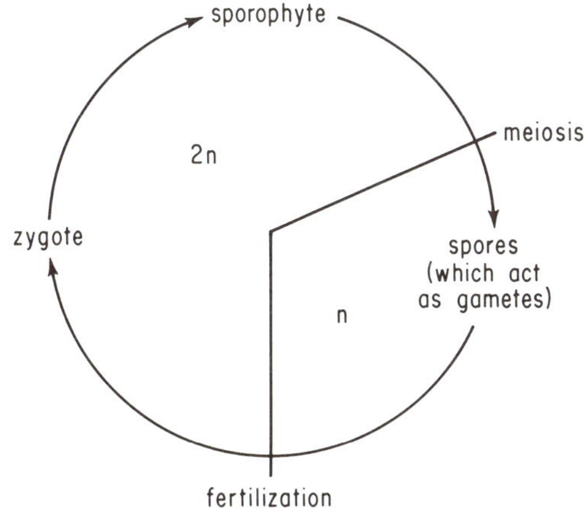

Fig. 2. Summary of a Life Cycle with Only a Sporophyte Generation

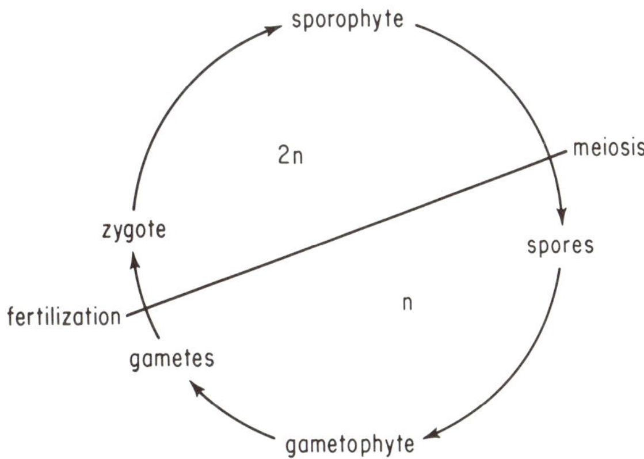

Fig. 3. Summary of a Life Cycle with Alternation of Generations

nomenclature two species of plants may not have the same binomial. Binomials are accepted by biologists everywhere. The binomial system was originated by CARL LINNAEUS, a Swedish naturalist, whose descriptive work, *Species Plantarum*, was published in 1753.

In taxonomic work the name (or an accepted abbreviation of it) of the person who first published a description of the species is added to the binomial. Thus the correct scientific name for the tree commonly called American elm in the United States is *Ulmus americana* L., the L. being an abbreviation of Linnaeus, who named this species. A binomial is always italicized, and the generic name is always capitalized. The species name usually is not capitalized though it may be if it is derived from a proper noun.

Although rules exist for the classifying of plants, not all plant taxonomists agree on what is the proper classification. Effort is made to have the classification reflect the evolutionary relationships of the taxa. Since there is not enough evidence to indicate clearly the lines of descent of all taxa, there is some difference of opinion among botanists, and different textbooks use slightly different systems of classification. Classifications are modified as additional information is obtained. The criteria used in classifying plants vary from group to group. Plants with similar life cycles are usually grouped together. In general more reliance is given to reproductive features than vegetative or biochemical ones, although all are used.

reproduction. The zygote is always the first cell of the sporophyte generation; the first cell of a gametophyte is always a spore.

In some species one or both generations have the ability to reproduce themselves asexually. Sometimes this is accomplished by rooting or by budding, but in other cases ASEXUAL SPORES are produced. These have the same number of chromosomes as the generation that produced them, and when they germinate they produce a generation identical to the one from which they arose. The spores produced by meiosis are called MEIOSPORES or SEXUAL SPORES.

Many variations are known to occur on these basic life cycles. The trend in several evolutionary lines has been toward dominance of the sporophyte generation. In the life cycles of a few species of algae and fungi more than two generations are known.

Most botanists divide the plant kingdom into two subkingdoms: Thallophyta and Embryophyta. Members of the THALLOPHYTA have reproductive organs that are derived from single cells and that are not surrounded by sterile tissue. Zygotes are released from the female organs and develop into mature plants without protection of the parent plant. Thallophytes have no vascular tissues and no true roots, stems, or leaves. Members of the EMBRYOPHYTA have multicellular reproductive organs surrounded by a layer of sterile tissue. Zygotes develop into embryos while still retained within the female reproductive organs. With a few exceptions the plants have true roots, stems, and leaves with vascular tissue.

The subkingdom THALLOPHYTA is usually divided into nine or ten divisions that include the algae and the fungi. All ALGAE contain the green pigment chlorophyll a, but members of the several divisions contain different accessory photosynthetic pigments that give many of the plants colors other than green. FUNGI lack chlorophyll and thus are unable to manufacture food. They are either parasites, obtaining their food from living organisms, or saprophytes, obtaining their food from dead organisms or from the waste products of living ones. In a broad sense the term fungi includes the bacteria, slime molds, and true fungi. LICHENS are a special group of thallophytes. A lichen is actually two organisms, a fungus and an alga living together symbiotically.

The subkingdom EMBRYOPHYTA has two divisions, the Bryophyta and Tracheophyta. The BRYOPHYTES (mosses and liverworts) lack true vascular tissue, and thus they have no true roots, stems, or leaves. The sporophyte generation is dependent on the gametophyte generation for food, water, and minerals. The gametophytes of some bryophytes superficially resemble stems bearing leaves. TRACHEOPHYTES have true roots, stems, and leaves with vascular tissue. The sporophyte is always an independent generation. The gametophyte is independent in some species; in others it is dependent on the sporophyte.

17 ALGAE

SELF-TEST

1. The algae are considered to be
   a a natural group of plants
   b an artificial group of plants
   c a group of plants showing clear evolutionary relationships
   d a group of plants having no features in common

2. The body of any alga is called a
   a filament
   b membrane
   c thallus
   d unicell

3. A flagellated asexual spore is called a
   a meiospore
   b oospore
   c zoospore
   d zygospore

4. The female reproductive organ of most algae is called a(n)
   a oogonium
   b archegonium
   c trichogyne
   d embryo sac

5. Two pigments common to all groups of algae are
   a chlorophyll a and chlorophyll b
   b chlorophyll b and phycocyanin
   c phycocyanin and carotene
   d carotene and chlorophyll a

6. The algae that synthesize most of the food utilized by aquatic animals are the
   a Rhodophyta, Phaeophyta, and Pyrrophyta
   b Euglenophyta, Chrysophyta, and Phaeophyta
   c Chrysophyta, Pyrrophyta, and Chlorophyta
   d Chlorophyta, Cyanophyta, and Rhodophyta

7. Pigments common to the blue-green algae and red algae but to no other algae are
   a phycocyanin and phycoerythrin
   b phycoerythrin and fucoxanthin
   c fucoxanthin and chlorophyll b
   d chlorophyll b and phycocyanin

8. Blue-green algae differ from green algae in having
   a sexual reproduction
   b no definite nucleus
   c flagella
   d chloroplasts

9. Algae having cell walls which contain silica and which are comprised of two separate overlapping halves are
   a red algae
   b blue-green algae
   c dinoflagellates
   d diatoms

10. The algae that are considered to be most closely related to the Embryophyta are the
    a blue-green algae
    b green algae
    c brown algae
    d red algae

1 ______
2 ______
3 ______
4 ______
5 ______
6 ______
7 ______
8 ______
9 ______
10 ______

BASIC FACTS

The algae are a diverse, widely distributed group, but they have certain features in common. Like the fungi they have unicellular reproductive organs that are not surrounded by sterile tissue. No embryos are formed, the zygotes being released from female reproductive organs and developing externally. The plant body, called a THALLUS (plural, thalli), does not have vascular tissue and is not differentiated into roots, stems, or leaves. The thallus may consist of a SINGLE CELL or a multicellular COLONY which may be FILAMENTOUS or MEMBRANOUS. A few algae are COENOCYTIC, the plant consisting of one or several large multinucleate cells.

Algae differ from the fungi in being AUTOTROPHIC, that is, they are capable of manufacturing their own food and are not dependent on other organisms. All algae have chlorophyll a and at least one other photosynthetic pigment that may give them a color other than green.

Algae may REPRODUCE sexually, asexually, or by both means. ASEXUAL REPRODUCTION is by fragmentation, asexual spores, or, in the case of unicellular plants, cell division. Flagellated, motile asexual spores are called ZOOSPORES. SEXUAL REPRODUCTION is of two types: isogamy and heterogamy. In ISOGAMY, the gametes are identical in appearance (isogametes); in HETEROGAMY, they are unlike in appearance (heterogametes), the egg usually being large and nonmotile and the sperm small and motile. Eggs are produced in a female reproductive organ, the OOGONIUM, and sperms in a male reproductive organ, the ANTHERIDIUM. The fusion of two isogametes is called CONJUGATION. The zygote is often a resistant spore that remains dormant over winter or during other unfavorable conditions. Such a zygote resulting from the fusion of two isogametes is called a ZYGOSPORE; that resulting from the union of two heterogametes is called an OOSPORE.

(Continued on page 68)

ADDITIONAL INFORMATION

The BLUE-GREEN ALGAE differ from other algae in having no true nucleus, no chloroplasts or mitochondria, and no sexual reproduction. Genetic material is present in a CENTRAL BODY that has no bounding membrane. When cell division occurs there is no mitosis. Chlorophyll is present on lamellae in the cytoplasm. The typical blue-green color of the plants is due to the presence of chlorophyll a and PHYCOCYANIN, a blue pigment. Carotene, xanthophylls, and PHYCOERYTHRIN (a red pigment) are sometimes present in large enough quantities to give other colors to the plants. Blue-green algae are single-celled, often occurring in colonies enclosed in GELATINOUS SHEATHS. They do not possess flagella and are therefore nonmotile, although a few species move by gliding or oscillating movements. Food is stored primarily as a glycogenlike carbohydrate. Blue-green algae are widely distributed, though most species grow in fresh water.

The GREEN ALGAE are a diverse group of chiefly freshwater plants ranging from unicellular to multicellular forms. Nearly all species reproduce sexually. A variety of types of life cycles, isomorphic and heteromorphic alternation of generations, and isogamy and heterogamy occur in this group. Green algae have definite nuclei. They have one or more chloroplasts containing the pigments chlorophyll a, chlorophyll b, carotene, and xanthophylls, which are the same pigments that are found in the Embryophyta. The chloroplasts often have PYRENOIDS, which are proteinaceous bodies surrounded by stored starch. Motile cells have two or four flagella at the anterior end.

The EUGLENOIDS are a small group of unicellular, freshwater algae that have no cell walls but a fairly rigid outer cytoplasm (PERIPLAST). One or more flagella arise from the base of an anterior GULLET. The chloroplasts have pigments similar to those of green algae. Although autotrophic, euglenoids become colorless and saprophytic if grown in the dark; some cells even ingest food particles through the gullet. Food is stored chiefly as PARAMYLUM, a starchlike carbohydrate. Sexual reproduction is rare. Asexual reproduction is by mitotic cell division. Euglenoids are often included in the animal kingdom.

The PYRROPHYTA are unicellular, mostly flagellated, marine algae. Plastids contain pigments including chlorophyll a, chlorophyll c, carotene, and several xanthophylls that make the plants yellow-brown, or occasionally red. Some species lack cell walls; others have cell walls containing a transverse and a longitudinal groove, each bearing a flagellum. Food is stored as starch or oil. Reproduction is by cell division. The dinoflagellates are also often considered members of the animal kingdom.

(Continued on page 68)

EXPLANATIONS

1. *The algae are considered to be* an artificial group of distantly related plants. This is shown by the fact that the algae encompass seven divisions. They have many different body forms, metabolisms, and types of reproduction.

2. *The body of any alga is called a* thallus, a plant body that is not differentiated into true roots, stems, and leaves and that lacks vascular tissue. It may take the form of a single cell, a membrane, or a filament.

3. *A flagellated asexual spore is called a* zoospore. Both sexual and asexual spores may be either motile or nonmotile.

4. *The female reproductive organ of most algae is called an* oogonium. In the red algae the female reproductive organ is called a carpogonium.

5. *Two pigments common to all groups of algae are* carotene and chlorophyll a. These pigments are common to all photosynthetic plants. Associated with these pigments are an assortment of xanthophylls and usually another chlorophyll (b, c, d, or e).

6. *The algae that synthesize most of the food utilized by aquatic animals are the* Chrysophyta, Pyrrophyta, and Chlorophyta. These algae form the base of food chains in aquatic habitats.

7. *Pigments common to the blue-green algae and red algae but to no other algae are* phycocyanin and phycoerythrin. The presence of these two pigments in both divisions and the fact that no motile cells occur in either division indicate that the blue-green and red algae may be related.

8. *Blue-green algae differ from green algae in having* no definite nucleus. They are not known to have sexual reproduction, flagella, or chloroplasts.

9. *Algae having cell walls which contain silica and which are comprised of two separate overlapping halves are* diatoms. Because of the silica, the cell walls do not decompose after death as the protoplasts do. Accumulations of these cell walls are known as diatomaceous earth. Diatomaceous earth is mined commercially and has several industrial uses.

10. *The algae that are considered to be most closely related to the Embryophyta are the* green algae, which have the same chloroplast pigments and food reserves; some also have alternation of generations. Some botanists feel that the Embryophyta evolved from green algae having isomorphic alternation of generations.

Answers

b	1
c	2
c	3
a	4
d	5
c	6
a	7
b	8
d	9
b	10

In many algae the thallus belongs to the GAMETOPHYTE GENERATION and produces haploid gametes, the zygote being the only diploid cell. Other algae have ALTERNATION OF GENERATIONS, in which both the gametophyte and the sporophyte generations undergo a distinct development. In a few algae the thallus belongs to the SPOROPHYTE GENERATION, and the gametes are the only haploid cells. Some red algae have three generations to the life cycle.

Algae are typically aquatic plants living in oceans, lakes, rivers, and temporary ponds. Most of the large algae and some of the smaller ones grow attached to submerged rocks by rootlike HOLDFASTS. Small algae and small aquatic animals that float freely on lakes or oceans are called PLANKTON. A few species of algae grow on soil or tree trunks, but even these require considerable moisture for growth.

Algae are CLASSIFIED on the basis of several features: reproduction, pigments, food reserves, motility, and presence or absence of a nucleus. There are seven fairly distinct groups of algae that are usually given the taxonomic rank of division: CYANOPHYTA (blue-green algae), CHLOROPHYTA (green algae), EUGLENOPHYTA (euglenoids), PYRROPHYTA (dinoflagellates and cryptomonads), CHRYSOPHYTA (diatoms, yellow-green algae, and golden-brown algae), PHAEOPHYTA (brown algae), and RHODOPHYTA (red algae).

The CHRYSOPHYTA are a large group of aquatic algae that include three classes: BACILLARIOPHYCEAE (diatoms), XANTHOPHYCEAE (yellow-green algae), and CHRYSOPHYCEAE (golden-brown algae). The diatoms are single-celled or filamentous. They usually have two large chloroplasts with chlorophyll a, chlorophyll c, carotene, and several xanthophylls that give a brown color. Oils are the major stored food. Cell walls are rich in silica. Each cell wall is composed of two overlapping halves called VALVES. Motile diatoms have slits, called RAPHES, through which streaming cytoplasm makes contact with water. Diatoms reproduce both sexually and asexually. The yellow-green algae have chlorophyll a, chlorophyll e, carotene, and an unidentified xanthophyll. They are unicellular (flagellated or amoeboid), coenocytic, or filamentous.

Most BROWN algae are macroscopic marine algae growing primarily in intertidal zones in cool regions of the world. The plant body is a branched filament, the branches sometimes so closely intertwined as to form structures that resemble the organs of higher plants. Chloroplasts contain chlorophyll a, chlorophyll c, carotene, and FUCOXANTHIN (a brown pigment) and other xanthophylls. Food is stored as LAMINARIN (a soluble polysaccharide) and MANNITOL (an alcohol). Motile cells (zoospores, isogametes, and sperms) have two lateral flagella. Sexual reproduction is both isogamous and heterogamous. Alternation of generations is either isomorphic or heteromorphic, and in some species there is only a sporophyte generation. Where heteromorphic alternation of generations occurs, the sporophyte is the dominant generation.

The RED ALGAE are chiefly multicellular, macroscopic marine forms living in warm waters. Plants may be membranous or filamentous, often finely branched and feathery. Stored food is FLORIDEAN STARCH, an insoluble carbohydrate. Chloroplasts contain chlorophyll a, chlorophyll d, carotene, a xanthophyll, phycocyanin, and PHYCOERYTHRIN, which makes the plants red. No motile cells are produced. Red algae have a specialized form of reproduction. Male gametes, called SPERMATIA, are produced in SPERMATANGIA. A spermatium is carried by water currents to the female reproductive organ, the CARPOGONIUM, which is similar to an oogonium but bears an extension called a TRICHOGYNE. The spermatium nucleus enters the carpogonium through the trichogyne. The zygote develops branched filaments bearing CARPOSPORES.

18 BACTERIA AND VIRUSES

SELF-TEST

1. A coccus is a(n)
   a rod-shaped cell
   b spherical cell
   c chain of cells
   d irregular clump of cells

2. Most bacteria differ from most other plants in that they lack
   a ribosomes, mitochondria, and nuclei
   b chloroplasts, mitochondria, and nuclei
   c DNA, cell walls, and chloroplasts
   d DNA, ribosomes, and cell walls

3. Bacteria reproduce
   a only sexually
   b only asexually
   c mostly sexually
   d mostly asexually

4. The nutrition of most bacteria is
   a heterotrophic, but that of a few species is photosynthetic or chemosynthetic
   b photosynthetic, but that of a few species is chemosynthetic
   c chemosynthetic, but that of a few species is photosynthetic or heterotrophic
   d autotrophic, but that of a few species is heterotrophic

5. The main role that bacteria play in the carbon cycle is
   a photosynthesis of organic compounds
   b chemosynthesis of organic compounds
   c digestion and oxidation of organic compounds
   d formation of macromolecules by condensation

6. Nitrifying bacteria
   a reduce nitrates to free nitrogen
   b oxidize ammonia to nitrates
   c convert free nitrogen to nitrogen compounds
   d convert proteins into ammonia

7. Some diseases caused by bacteria are
   a tetanus, typhoid, and tuberculosis
   b measles, malaria, and mumps
   c smallpox, sleeping sickness, and syphilis
   d pneumonia, poliomyelitis, and psittacosis

8. Viruses differ from other living things in having no
   a protein
   b nucleic acids
   c cellular organization
   d method of utilizing energy

9. Most viruses are composed of
   a nucleic acid and cellulose
   b cellulose and fats
   c fats and protein
   d protein and nucleic acid

10. A substance that is produced by one species and that inhibits the growth of members of another species is an
    a antiseptic
    b antibiotic
    c antigen
    d anticoagulant

1 ______
2 ______
3 ______
4 ______
5 ______
6 ______
7 ______
8 ______
9 ______
10 ______

BASIC FACTS

BACTERIA are usually classified in the SCHIZOMYCOPHYTA, one of the divisions of fungi. Because bacteria reproduce asexually by fission and lack nuclei and chloroplasts, they and the blue-green algae, which share these characteristics, sometimes are classified as two classes in the same division. Bacteria are microscopic, single-celled organisms. Cells are smaller than those of other organisms, diameters usually ranging from 0.5μ to 1.5μ. There are three SHAPES of bacterial cells: the spherical cell called a COCCUS, the rod-shaped cell called a BACILLUS, and the spiral-shaped cell called a SPIRILLUM. In some species a few cells stay together in clumps or chains. Cocci that remain in pairs are DIPLOCOCCI; those that remain in chains are STREPTOCOCCI; those in irregular clumps are STAPHYLOCOCCI; and a cubical packet of eight is a SARCINA.

Bacterial cells have CELL WALLS composed of various compounds, and the cells are surrounded by a SLIME LAYER, or CAPSULE. Some bacteria are motile, most of them by FLAGELLA. Bacteria lack mitochondria as well as chloroplasts and true nuclei. Their genetic material occupies a CENTRAL BODY of the cell, but it is not surrounded by a membrane. REPRODUCTION is largely by fission. The reproduction rate is high; under optimum conditions a generation time may be as short as twenty minutes. A modified type of sexual reproduction (called conjugation) occurs in some bacteria. Some species form SPORES, resistant structures that can survive unfavorable conditions. Spores are not reproductive structures, for only one spore is produced by a vegetative cell.

Most bacteria are HETEROTROPHIC, that is, they are dependent on other organisms for their food. Some are saprophytes, others parasites. Many of the latter are pathogens (disease-

(Continued on page 72)

ADDITIONAL INFORMATION

Bacteria OCCUR in almost every conceivable habitat. They are present in natural waters and in soils. They are carried by air currents throughout the world. They are part of the natural flora of the digestive tracts of animals. They are rare in very dry areas, and their numbers become fewer with increasing depth in soil.

Bacteria vary in their oxygen relationships. OBLIGATE AEROBES are those which require atmospheric oxygen. OBLIGATE ANAEROBES cannot grow in the presence of free oxygen. Most bacteria are FACULTATIVE ANAEROBES; they can utilize free oxygen but do not require it.

The small size of bacteria gives them a great deal of surface area per volume. Since their metabolism depends on the movement of materials across the cell membranes, the large surface area allows for rapid rates of chemical reactions. For this reason bacteria quickly cause disease, food spoilage, or decomposition of dead plants and animals. Some bacteria-caused DISEASES of man are food poisoning, pneumonia, tetanus, typhoid, plague, tuberculosis, syphilis, and bacillary dysentery. Bacteria and viruses are spread from one host to another by several means. Those causing respiratory diseases are often carried through the air in droplets discharged by sneezing or coughing. A new infection can occur when these droplets are inhaled. Intestinal diseases are often spread by contaminated water supplies and by foods that have come in contact with such water or that have been handled by diseased persons. Parasites that live in the blood are usually transmitted by animal bites. Bacteria and viruses can also be spread by direct contact.

Because of their biochemical activities, bacteria and true fungi are used COMMERCIALLY in the production of foods, antibiotics, and other materials. In the souring of milk they convert lactose to lactic acid. Ethyl alcohol is oxidized to acetic acid in the production of vinegar from wine or hard cider.

ACTINOMYCETES are funguslike soil bacteria that tend to form branched filaments; they reproduce asexually by the formation of conidiospores. Some actinomycetes are the source of ANTIBIOTICS, substances which are produced by organisms and which inhibit the growth of other organisms. Antibiotics are used in fighting many bacterial and some viral infections in humans and animals.

Bacteria help to complete some of the great natural cycles. In the CARBON CYCLE, autotrophic plants convert carbon dioxide in the air to an organic form by photosynthesis and to a lesser extent by chemosynthesis. When plants are eaten by animals, these carbon compounds are utilized by the animals in building up their own tissues. Some of this organic carbon is oxidized to carbon dioxide by the respiration

(Continued on page 72)

EXPLANATIONS

1. *A coccus is a* spherical cell. This is one of the three shapes of bacterial cells.

2. *Most bacteria differ from most other plants in that they lack* chloroplasts, mitochondria, and nuclei. Like blue-green algae, bacteria have genetic material confined to a central area that is not a true nucleus, and those bacteria that are photosynthetic have chlorophyll present on cytoplasmic lamellae rather than in organized chloroplasts.

3. *Bacteria reproduce* mostly asexually. In those species in which a modified form of sexual reproduction has been demonstrated, only a few out of millions of cells reproduce this way.

4. *The nutrition of most bacteria is* heterotrophic, but that of a few species is photosynthetic or chemosynthetic. The latter obtain energy from inorganic oxidations.

5. *The main role that bacteria play in the carbon cycle is* digestion and oxidation of organic compounds to carbon dioxide. These reactions prevent the accumulation of waste products and dead organisms as well as returning carbon dioxide to the air.

6. *Nitrifying bacteria* oxidize ammonia to nitrates; this process is called nitrification.

7. *Some diseases caused by bacteria are* tetanus, typhoid, and tuberculosis; also syphilis, and pneumonia.

8. *Viruses differ from other living things in having no* cellular organization. Viruses are obligate parasites because they require living hosts as an energy source.

9. *Most viruses are composed of* protein and nucleic acid. Some of the larger viruses contain other compounds.

10. *A substance that is produced by one species and that inhibits the growth of members of another species is an* antibiotic. Bacteria and the true fungi produce most of the antibiotics that are useful in medicine.

Answers

Answer	No.
b	1
b	2
d	3
a	4
c	5
b	6
a	7
c	8
d	9
b	10

causing organisms). A few species are AUTOTROPHIC; some of these are photosynthetic, and others are chemosynthetic. BACTERIAL PHOTOSYNTHESIS differs from photosynthesis in other plants in that water is not the source of hydrogen; ammonia, hydrogen sulfide, gaseous hydrogen, or other compounds are the hydrogen donors. Bacterial chlorophyll resembles, but is not identical with, other chlorophyll. CHEMOSYNTHETIC BACTERIA manufacture food with the energy released by the oxidation of inorganic compounds. These bacteria are not dependent on light.

Although bacteria show very little morphological diversity, they are capable of carrying on many different METABOLIC PROCESSES. In the process of DECAY, bacteria convert the compounds present in dead plants and animals or their waste products into compounds that can be used by other organisms. FERMENTATION oxidizes sugars to simpler organic compounds such as ethyl alcohol or lactic acid. Decay ultimately converts carbohydrates, fats, and proteins into inorganic compounds: carbon dioxide, water, ammonia, nitrates, hydrogen sulfide, and sulfates. A few bacteria are NITROGEN FIXERS and convert gaseous nitrogen of the air into a form usable by green plants.

VIRUSES are acellular, ultramicroscopic organisms consisting of a nucleic acid core covered by a protein coat. The genetic material is either DNA or RNA. Viruses are all parasites. When they are inside a host they exhibit properties of living things, but they may be removed and crystallized. The crystals may be stored indefinitely, but when injected into a suitable host, they resume viral activity.

processes of the plant or animal in which they occur, but a good deal of it remains in the reduced form when the organism dies. DECAY BACTERIA and other heterotrophic microorganisms digest these organic compounds and then, in their own respiration, oxidize them to carbon dioxide which returns to the air, thus completing the cycle. All carbon compounds in the cycle can be used by some organisms. Carbon is continually being recycled by the activities of green plants, animals, and microorganisms.

In the NITROGEN CYCLE, plants absorb nitrogen from the soil in the form of nitrates and convert it to amino acids, which are the building blocks of proteins. When a plant is eaten by an animal, these proteins are utilized by the animal in building up its own body proteins. After plants and animals die, DECAY BACTERIA digest the proteins of which they are composed, first to amino acids and then to ammonia. NITRIFYING BACTERIA oxidize ammonia to nitrites and nitrites to nitrates which plants can absorb, thus completing the cycle. Under anaerobic conditions, DENITRIFYING BACTERIA reduce nitrates to nitrites and nitrites to free nitrogen (N_2) which cannot be used by most plants. A few bacteria and blue-green algae are NITROGEN FIXERS; they convert free nitrogen to a reduced organic form usable by higher plants. *Rhizobium* is a symbiotic nitrogen-fixing bacterium that grows in the nodules of leguminous plants. Free-living nitrogen-fixing bacteria and blue-green algae add fixed nitrogen to the soil.

VIRUSES lack many enzymes and are dependent on the enzyme system of a host cell. When a virus infects a host cell, only the nucleic acid enters the cell. The nucleic acid utilizes the food and enzymes of the host to synthesize additional viruses. Within a half hour after infection by a single virus more than a hundred new viruses can be produced. Viruses infect both plants and animals. Some virus-caused diseases of humans are smallpox, yellow fever, rabies, measles, poliomyelitis, and the common cold. Viruses that infect bacteria are called BACTERIOPHAGES. They are tadpole-shaped, having a tail and an enlarged head region containing DNA. The tail attaches to the bacterial cell and the DNA passes through its hollow core into the protoplast, leaving the protein coat of the head behind.

19 SLIME MOLDS AND TRUE FUNGI

SELF-TEST

1. The fungi that most resemble animals are
   a Myxomycophyta
   b Ascomycetes
   c Basidiomycetes
   d Phycomycetes

2. An aggregation of single ameboid cells of slime molds is called a
   a gametangium
   b plasmodium
   c sporangium
   d mycelium

3. The coenocytic condition is typical of
   a lichens
   b Basidiomycetes
   c Ascomycetes
   d Phycomycetes

4. In their reproduction the Phycomycetes most closely resemble
   a club fungi
   b imperfect fungi
   c mosses
   d algae

5. Dicaryon mycelia are characteristic of
   a Ascomycetes and slime molds
   b Ascomycetes and Basidiomycetes
   c Phycomycetes and slime molds
   d Phycomycetes and Basidiomycetes

6. Karyogamy is
   a the fusion of gamete nuclei
   b the fusion of gamete protoplasts
   c delayed meiosis
   d delayed mitosis

7. Sexual spores in fungus life cycles include
   a conidiospores and swarm spores
   b swarm spores and ascospores
   c ascospores and basidiospores
   d basidiospores and conidiospores

8. The life cycle of a typical Ascomycete or Basidiomycete can be represented by
   a $n \rightarrow n + n \rightarrow n$
   b $n \rightarrow 2n \rightarrow n$
   c $n \rightarrow n + n \rightarrow 2n \rightarrow n$
   d $n \rightarrow 2n \rightarrow n + n \rightarrow n$

9. A basidiocarp may be a
   a water mold
   b mushroom
   c yeast
   d truffle

10. The gametes of typical Basidiomycetes are morphologically indistinguishable from
    a gametes of the Ascomycetes
    b basidiospores
    c vegetative cells
    d gametes of the Phycomycetes

11. Imperfect fungi lack
    a spores
    b sexual reproduction
    c asexual reproduction
    d hyphae

12. Most lichens are composed of a
    a green alga and an Ascomycete
    b green alga and a Basidiomycete
    c blue-green alga and an Ascomycete
    d blue-green alga and a Basidiomycete

13. Teliospores are found in
    a rusts
    b yeasts
    c molds
    d mildew

1 ______
2 ______
3 ______
4 ______
5 ______
6 ______
7 ______
8 ______
9 ______
10 ______
11 ______
12 ______
13 ______

BASIC FACTS

As thallophytes, the SLIME MOLDS and the TRUE FUNGI share with the algae and the bacteria certain characteristics: unicellular reproductive organs, no embryo formation, and a plant body that is a THALLUS. Unlike algae, fungi have no chlorophyll and are heterotrophic, either saprophytic or parasitic.

The SLIME MOLDS (division MYXOMYCOPHYTA) have both plant and animal characteristics. During the vegetative phase of the life cycle the cells have no walls and move about in ameboid fashion either singly or aggregated into a mass called a PLASMODIUM. These stages may engulf whole food. From the plasmodium develops a reproductive stage called a SPORANGIUM. It has plantlike features: it is attached to a substrate and its cells have walls.

The thallus of most TRUE FUNGI (division EUMYCOPHYTA) is a filament called a HYPHA. Hyphae often branch profusely, the entire branching system being called a MYCELIUM. In some fungi the hyphae form large FRUITING BODIES. True fungi have cell walls that contain CHITIN. There are four CLASSES of true fungi: the algalike fungi (PHYCOMYCETES), sac fungi (ASCOMYCETES), club fungi (BASIDIOMYCETES), and imperfect fungi (FUNGI IMPERFECTI or DEUTEROMYCETES). The sexual phase of a fungus life cycle is called the PERFECT STAGE, the asexual phase the IMPERFECT STAGE.

Most species of PHYCOMYCETES are coenocytic, the hyphae, which are tubular, having no cross walls except those that separate reproductive organs from the rest of the mycelium. Some species produce flagellated zoospores; Phycomycetes is the only class of true fungi having any members with motile spores. There are no definite fruiting bodies. Life cycles of the Phycomycetes are similar to those of

(Continued on page 76)

ADDITIONAL INFORMATION

SLIME MOLD spores germinate into ameboid cells lacking walls. The cells move about, feeding on the substrate and dividing several times. Eventually single cells act as gametes and fuse in pairs forming zygotes. A plasmodium is produced either by fusion of zygotes or by nuclear divisions of the zygote unaccompanied by cytokinesis. The naked plasmodium moves as a single unit, feeding on rotting vegetation. Later the plasmodium is converted into sporangia that rise above the substrate. Meiosis precedes the formation of spores in the sporangia, so the spores and gametes are haploid.

Primitive members of the PHYCOMYCETES are aquatic. In many species gametes and spores are motile by flagella; in a few (such as *Saprolegnia*) the male gametes are merely nuclei produced in an antheridium which discharges them directly into the oogonium. The more advanced Phycomycetes, like *Rhizopus*, the common bread mold, are terrestrial with no motile cells. Sexual reproductive structures of *Rhizopus* are multinucleate cells called GAMETANGIA. They are produced on short hyphae, called SUSPENSORS, when mycelia of opposite mating types (plus and minus) are present. After the hyphae make contact with each other, cell walls form, separating each gametangium from its hypha. Conjugation of two gametangia produces a ZYGOSPORE with many diploid nuclei. A hard wall forms around the zygospore, which enters a dormant period. The nuclei undergo meiosis before the zygospore germinates into a short, sporangium-bearing mycelium.

ASCOMYCETES and BASIDIOMYCETES are terrestrial fungi with no motile cells. Their REPRODUCTION is peculiar in that the fusion of gamete nuclei usually does not immediately follow fusion of protoplasts of the gametes. Fusion of protoplasts is called PLASMOGAMY; fusion of nuclei is called KARYOGAMY. Plasmogamy produces a dicaryon cell. When this cell divides each nucleus divides also, a daughter cell receiving one nucleus from each resulting pair. Repeated divisions produce DICARYON HYPHAE, each cell of which has two nuclei. (Dicaryon cells are not actually diploid, for each nucleus is haploid; the chromosome number is usually indicated as $n + n$.) The dicaryon hyphae form the fruiting bodies (ascocarps). Within each fruiting body is a fertile area, the HYMENIUM. Karyogamy occurs within one cell near the tip of most dicaryon hyphae in the hymenium. This cell is a zygote and is the only diploid cell in the life cycle. It is called an ASCUS (in Ascomycetes) or a BASIDIUM (in Basidiomycetes). The zygote nucleus undergoes meiosis, producing four haploid nuclei within this cell. In Ascomycetes the nuclei then undergo one mitotic division, producing eight haploid nuclei around which cell walls develop; each nucleus with its cytoplasm and cell wall is an ASCOSPORE. In Basidiomycetes the four nuclei usually migrate to the tip of the

(Continued on page 76)

EXPLANATIONS

1. *The fungi that most resemble animals are* Myxomycophyta (slime molds), which have an ameboid stage in the life cycle and which ingest whole food.

2. *An aggregation of single ameboid cells of slime molds is called a* plasmodium, the feeding stage in the life cycle.

3. *The coenocytic condition is typical of* Phycomycetes. Other members of the Eumycophyta have cross walls dividing the mycelia into cells.

4. *In their reproduction the Phycomycetes most closely resemble* algae, for which they are named.

5. *Dicaryon mycelia are characteristic of* Ascomycetes and Basidiomycetes. Except for lichens, no other groups of plants have dicaryon mycelia.

6. *Karyogamy is* the fusion of gamete nuclei. It occurs after plasmogamy, the fusion of the protoplasts of two gametes. Karyogamy may follow plasmogamy immediately or it may be delayed.

7. *Sexual spores in fungus life cycles include* ascospores and basidiospores. They are produced by meiotic divisions in asci and basidia. Conidiospores are asexual spores.

8. *The life cycle of a typical Ascomycete or Basidiomycete can be represented by* $n \rightarrow n + n \rightarrow 2n \rightarrow n$. Plasmogamy of haploid gametangia or gametes (n) results in dikaryon mycelia ($n + n$). Karyogamy produces a diploid zygote ($2n$) which by meiosis produces haploid sexual spores (n). The spores germinate into haploid, monocaryon mycelia on which are formed the reproductive organs.

9. *A basidiocarp may be a* mushroom. Basidiocarps are fruiting bodies of some basidiomycetes.

10. *The gametes of typical Basidiomycetes are morphologically indistinguishable from* vegetative cells. Any cell of a monocaryon mycelium can act as a gamete.

11. *Imperfect fungi lack* sexual reproduction. Most imperfect fungi reproduce by asexual spores; a few reproduce by fragmentation of the mycelium.

12. *Most lichens are composed of a* green alga and an Ascomycete. A few lichens have other algae or other fungi as members.

13. *Teliospores are found in* rusts and smuts.

Answers

a	1
b	2
d	3
d	4
b	5
a	6
c	7
c	8
b	9
c	10
b	11
a	12
a	13

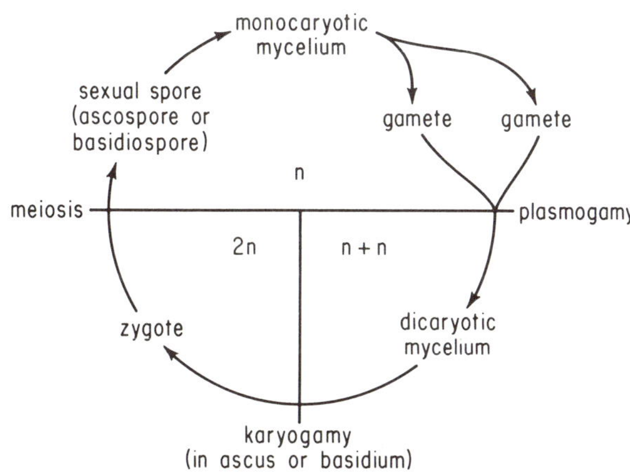

Summary of the Life Cycle of a Typical Ascomycete or Basidiomycete

many algae. Isogamy and heterogamy, and both sexual and asexual spores occur in this class. The bread mold fungus, water molds, and the downy mildews are members of the Phycomycetes.

The hyphae of ASCOMYCETES have cross walls that separate them into cells with one or two haploid nuclei. Most of the life cycle is MONOCARYOTIC (uninucleate). Sexual spores, called ASCOSPORES, are produced in a saclike structure at the end of a hypha called an ASCUS. Usually eight ascospores are formed in an ascus. Asci are frequently grouped in a fruiting body called an ASCOCARP. The female reproductive organ is called an ASCOGONIUM, the male an ANTHERIDIUM; both are multinucleate. CONIDIOSPORES are asexual spores that are usually produced terminally on special hyphae. Members of the Ascomycetes include yeasts, powdery mildews, and cup fungi.

The BASIDIOMYCETES also have hyphae with crosswalls. They have both MONOCARYOTIC and DICARYOTIC (binucleate) phases, the latter comprising most of the life cycle. There are no morphologically distinct sex organs; monocaryotic vegetative cells act as gametes. Sexual spores, called BASIDIOSPORES, are borne externally on a BASIDIUM. Usually four basidiospores are formed. Basidia are frequently produced in a conspicuous fruiting body called a BASIDIOCARP familiar to many persons as a mushroom, a puffball, or a bracket fungus.

The class FUNGI IMPERFECTI is an artificial group composed of those fungi whose perfect stage is absent or unknown.

LICHENS, usually classified with the fungi, consist of a fungus which has an alga living within it. The photosynthetic alga provides food to the fungus; the fungus absorbs water and affords protection to the alga. The fungus partner is usually an Ascomycete, the alga a single-celled green alga.

basidium and together with some cytoplasm are extruded from the basidium. Each of these four cells is a BASIDIOSPORE. The spores are usually forcibly discharged from their asci and basidia. Germinating ascospores and basidiospores produce MONOCARYOTIC HYPHAE.

Ascogonia and antheridia of the ASCOMYCETES are multinucleate and are produced on monocaryotic mycelia. The nuclei are all haploid. The ascogonium bears an elongated appendage, the TRICHOGYNE, which grows toward an antheridium and makes contact with it. Antheridial nuclei enter the trichogyne and travel to the main body of the ascogonium. Antheridial and ascogonial nuclei pair but do not fuse. Several filaments begin to form on the ascogonium; each receives a pair of nuclei. These dicaryon hyphae are the ascogenous hyphae that form the ascocarps. Three types of ASCOCARP are: the CLEISTOTHECIUM, which is spherical and has no opening; the PERITHECIUM, which is flask-shaped and has an opening; and the APOTHECIUM, which is cup-shaped. ASEXUAL REPRODUCTION is by any of several asexual spores, commonly by CONIDIOSPORES (CONIDIA), which are produced in chains on branched or unbranched stalks called CONIDIOPHORES.

The YEASTS are the unicellular Ascomycetes that form no ascocarps and have no dicaryon stages. They reproduce largely by BUDDING, a form of cell division in which one of the two daughter nuclei remains near the center of the parent cell and the other migrates toward the cell wall. A small bud develops at this point and the latter nucleus moves into it. The bud may detach from the parent cell immediately or it may remain until it has produced one or more of its own buds. Any yeast cell may act as a gamete. Fusion of two cells immediately produces a diploid zygote which usually acts as an ascus.

Some ASCOMYCETES cause diseases of economically important plants, but others are useful. *Pencillium* is the source of the antibiotic penicillin. Yeasts are used commercially because of their ability to ferment sugar to alcohol and carbon dioxide.

The RUSTS and SMUTS are BASIDIOMYCETES that do not produce large fruiting bodies. They are obligate parasites and do a great deal of damage to crop plants, particularly the cereal grains. Wheat rust (*Puccinia graminis*) has a complex life cycle with two different hosts and five separate stages, each producing its own kind of spore. SPERMAGONIA (or PYCNIA) with spermatia (or pycniospores) and AECIA with aeciospores occur on leaves of American barberry. UREDINIA with uredospores and TELIA with teliospores occur on stems of wheat. Teliospores produced on wheat fall to the ground, where they overwinter. In spring BASIDIA are produced directly on teliospores, and basidiospores reinfect the barberry.

20 BRYOPHYTES

SELF-TEST

1. Which of the following is *not* a characteristic of the Embryophyta:
   a unicellular reproductive organs
   b embryos nourished by gametophytes
   c food stored as starch
   d alternation of generations

2. Spores produced by the Bryophyta are
   a all asexual
   b all sexual
   c mostly asexual
   d mostly sexual

3. A venter is part of the
   a archegonium
   b antheridium
   c sporangium
   d sporophyte

4. In bryophytes the diploid number of chromosomes occurs in the
   a gametes
   b spores
   c nuclei of the gametophyte
   d spore mother cells

5. Bryophytes typically
   a have nonphotosynthetic gametophytes
   b have independent sporophytes
   c lack vascular tissue
   d are not dependent on water for fertilization

6. The calyptra and operculum are derived from the
   a sporophyte
   b gametophyte
   c gametophyte and sporophyte respectively
   d sporophyte and gametophyte respectively

7. Spore dissemination in some liverworts is aided by
   a the columella
   b peristome teeth
   c elaters
   d the calyptra

8. A germinating moss spore produces a(n)
   a protonema
   b apophysis
   c peristome
   d sporophyte

9. In the Bryophyta the zygote develops into an embryo in the
   a gemma
   b paraphyses
   c capsule
   d venter

10. The Anthocerotae differ from the other bryophytes in having
    a no alternation of generations
    b a gametophyte dependent on the sporophyte
    c a sporophyte with a persistent meristem
    d a sporophyte with no chlorophyll

1 ______
2 ______
3 ______
4 ______
5 ______
6 ______
7 ______
8 ______
9 ______
10 ______

BASIC FACTS

Members of the subkingdom EMBRYOPHYTA differ from those of the subkingdom Thallophyta in having multicellular reproductive organs the fertile cells of which are surrounded by a layer of sterile cells. The female reproductive organ is called an ARCHEGONIUM rather than an oogonium; within it the zygote develops into an EMBRYO that receives nourishment from the gametophyte. Members of the Embryophyta typically have alternation of a haploid gametophyte generation with a diploid sporophyte generation. At least one generation is photosynthetic; green cells usually have many small disc-shaped chloroplasts. The plastid pigments are the same as those of the green algae: chlorophyll a, chlorophyll b, carotene, and xanthophylls. Food is often stored as starch.

The subkingdom Embryophyta has two divisions: Bryophyta and Tracheophyta. BRYOPHYTES are small plants that lack vascular tissue; most are confined to moist, shady habitats. They often form dense mats on soil, rocks, rotting logs, or the bases of tree trunks. Bryophytes are typically terrestrial plants with rootlike, water-absorbing, anchoring organs called RHIZOIDS. An epidermis covered with cutin reduces transpiration, but the bryophytes vary in the amount of water they require to prevent desiccation; a few species are aquatic. GAMETOPHYTES grow attached to the substrate; they are photosynthetic and independent plants and are the dominant generation. An archegonium has only one egg, but an antheridium produces hundreds of sperms which are the only motile cells. Since sperms swim to the egg in a thin layer of water covering the surface of the plants, bryophytes are dependent on water for fertilization. The SPOROPHYTE is divided into three parts: a FOOT, which is embedded in the archegonium and absorbs food, water, and minerals from it; a SETA (stalk); and a CAPSULE (or sporangium), which produces many spores. All spores are

(Continued on page 80)

ADDITIONAL INFORMATION

An ARCHEGONIUM consists of two parts: a VENTER and a NECK. The venter contains one egg and a sterile VENTRAL CANAL CELL. In a mature archegonium the neck is hollow and open at the top. Sperms enter the archegonium through the neck canal. The ANTHERIDIA produce many biflagellate sperms surrounded by a layer of sterile cells. Both archegonia and antheridia generally are stalked.

Leafy LIVERWORTS and some thallose liverworts have little internal differentiation of tissues. In other thallose liverworts, such as *Ricciocarpus* and *Marchantia*, the gametophytes have an upper photosynthetic tissue and a lower storage tissue. The upper tissue is divided into chambers each of which opens to the air by a single pore through which exchange of carbon dioxide and oxygen occurs. Scales on the lower surface conserve moisture by holding water in capillary spaces.

The sporophyte of *Ricciocarpus* has no foot or stalk; it consists only of a CAPSULE that remains within the venter of the archegonium. The capsule wall encloses many spore mother cells which divide by meiosis and produce haploid spores. Spores are disseminated only after death and decay of the gametophyte.

In *Marchantia* the antheridia are borne upright in small chambers in the flat top of a stalk called an ANTHERIDIOPHORE. Sperms escape through small pores at the tops of the chambers and are disseminated by raindrops. The archegonia hang upside down between fingerlike lobes at the top of a stalk called an ARCHEGONIOPHORE. The sporophyte hangs upside down, its foot embedded in the venter. The spores are mixed with hygroscopic ELATERS, elongated cells that twist and untwist with changes in moisture content. When the capsule opens, the elaters aid in disseminating the spores. Spores germinate into gametophytes.

ASEXUAL REPRODUCTION of liverworts is by fragmentation of the gametophyte or by production of multicellular buds called GEMMAE. The gemmae of *Marchantia* are formed in GEMMAE CUPS on the upper surface of the thallus. Gemmae are disseminated when raindrops splash them out of the cups.

The HORNWORTS are similar to liverworts, but they have several differences that place them in a separate class. Cells of the gametophytes usually have one large chloroplast, those of the sporophyte two; the chloroplasts have pyrenoids and resemble the plastids of the algae. Both generations are photosynthetic. The epidermis of the sporophyte capsule contains STOMATA. The capsule is long and slender, opening by a longitudinal slit at the top. It is unusual in having a BASAL MERISTEM; the meristem continues to produce new tissues during the life of the sporophyte. Newly formed spore mother cells push older ones upward. The capsule has a central column of sterile cells called the COLUMELLA; as it dries and twists it assists in dissemination of spores.

(Continued on page 80)

EXPLANATIONS

1. Unicellular reproductive organs are not characteristic of the Embryophyta. The female reproductive organ is the archegonium, the male organ the antheridium. Embryophytes retain their embryos within the archegonium where they are nourished.

2. *Spores produced by the Bryophyta are* all sexual spores produced by the sporophyte. Asexual reproduction occurs, but it is by some means other than spores.

3. *A venter is part of the* archegonium; the venter contains an egg and is surmounted by a long neck. The single egg is the only fertile cell in the archegonium. The presence of sterile cells distinguishes the archegonium of the Embryophyta from the oogonium of the Thallophyta.

4. *In bryophytes the diploid number of chromosomes occurs in the* spore mother cells. In all members of the Embryophyta meiosis always occurs in spore mother cells located in the capsule (Bryophyta) or sporangium (Tracheophyta). Meiosis always precedes the formation of spores; it never precedes the formation of gametes.

5. *Bryophytes typically* lack vascular tissue. Gametophytes usually are small, green plants. Although they are independent, their lack of an extensive conducting system keeps them from reaching large sizes. Most members are not more than one or two inches high. Gametophytes supply sporophytes with water and minerals and usually with at least part of their food.

6. *The calyptra and operculum are derived from the* gametophyte and sporophyte respectively. The calyptra is part of the archegonium, and the operculum is part of the capsule. However, in the mature sporophyte the calyptra has become detached from the gametophyte and covers the capsule.

7. *Spore dissemination in some liverworts is aided by* elaters. Peristome teeth aid in spore dissemination in mosses. The columella plays a part in disseminating hornwort spores.

8. *A germinating moss spore produces a* protonema which gives rise to adult gametophytes.

9. *In the Bryophyta the zygote develops into an embryo in the* venter of the archegonium. The foot of the sporophyte remains in the venter as an absorbing organ during the entire life of the sporophyte.

10. *The Anthocerotae differ from the other bryophytes in having* a sporophyte with a persistent meristem that is active throughout the life of the sporophyte.

Answers

a	1
b	2
a	3
d	4
c	5
c	6
c	7
a	8
d	9
c	10

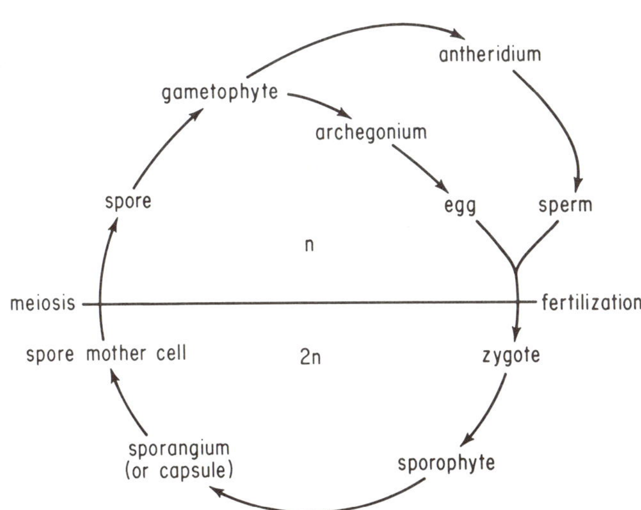

Summary of the Life Cycle of a Typical Embryophyte

sexual spores; they are thick-walled, nonmotile cells that are usually transported by wind.

The bryophytes are divided into three classes: Hepaticae (liverworts), Anthocerotae (hornworts), and Musci (mosses).

The LIVERWORTS have flattened, prostrate gametophytes that may take the form of a ribbon-shaped, dichotomously branching thallus or that may be differentiated into structures externally resembling leaves and stems. The lower surface of the gametophyte bears unicellular rhizoids that absorb water and minerals. In most species sex organs are produced on the upper surface and are protected from desiccation by the growth of sterile tissue around them. Sporophytes are dependent on the gametophytes for food, water, and minerals. *Ricciocarpus, Marchantia,* and *Porella* are liverworts.

The HORNWORTS have a flat, prostrate, thallose gametophyte with rhizoids on the lower surface and the sex organs embedded in chambers just below the upper surface. The slender sporophytes stand erect above the gametophytes. Sporophytes are photosynthetic and depend on the gametophytes only for water and minerals. The best-known hornwort is *Anthoceros.*

Gametophytes of MOSSES may be erect or prostrate; they are differentiated into stemlike and leaflike structures, and they have radial symmetry. Rhizoids at the base of the stems absorb water and minerals. Sex organs are produced at the tips of stems. The early, filamentous stage of the gametophyte is called the PROTONEMA. Young sporophytes are often green, but older sporophytes usually lack chlorophyll. They are dependent on gametophytes for food and water and probably for part of their food. Examples of mosses are *Mnium, Polytrichum, Funaria,* and *Sphagnum.*

Moss spores germinate into green, branched, algalike filaments called PROTONEMATA. Buds which appear along the length of a protonema give rise to adult LEAFY GAMETOPHYTES. The stems and leaves of gametophytes of some mosses have conducting tissue that superficially resembles vascular tissue. RHIZOIDS occur on the protonemata and on the lower portion of the adult gametophytes. The archegonia and antheridia form at the tips of the stems. Among them are short sterile hairs called PARAPHYSES that may conserve moisture around the sex organs.

The mature sporophyte capsule has a central sterile COLUMELLA surrounded by sporogenous tissue that in turn is surrounded by photosynthetic tissue. Stomata frequently occur in the epidermis of the lower part of the capsule. The capsule is covered by a CALYPTRA which is part of the archegonium in which it was formed; as the seta elongates, the archegonium tears off and is carried upward on the capsule. The capsule itself has a lid, or OPERCULUM. At maturity the capsule usually bends down and both calyptra and operculum are shed. Beneath the operculum is a set of appendages called PERISTOME TEETH. They are hygroscopic; in wet weather they close down over the mouth of the capsule and prevent the spores from dropping out, but in dry weather they bend back and let the spores escape. Dissemination in dry weather increases the chances of spores being blown great distances by wind.

Except for *Sphagnum*, the bog moss or PEAT MOSS, bryophytes have little economic value. *Sphagnum* grows in bogs with water so acid that decay is delayed, and over many years the dead moss accumulates in large deposits called peat; when dried it can be used as fuel. The leaves of the gametophytes have large dead cells that hold great quantities of water. Peat moss is used to improve the water-holding capacities of sandy soils and to pack plants for shipment.

Many bryophytes are important PIONEER PLANTS in newly colonized areas. They contribute to the formation of soil on which larger plants can grow. Dense mats of bryophytes retain water and thus prevent erosion.

The ANCESTRY of the bryophytes is uncertain, but many botanists think that they are descended from the earliest land plants. From these plants evolved two main lines: the Bryophyta with relatively simple plant bodies and the more complex Tracheophyta (vascular plants).

21 PSILOPSIDA, LYCOPSIDA, AND SPHENOPSIDA

SELF-TEST

1. All tracheophytes have
   a independent gametophytes
   b dependent gametophytes
   c independent sporophytes
   d dependent sporophytes

2. Two features that the living Psilopsida, Lycopsida, and Sphenopsida have in common are
   a microphylls and lack of seeds
   b lack of seeds and possession of vessels
   c possession of vessels and secondary growth
   d secondary growth and microphylls

3. Compared with the gametophytes of the bryophytes, the gametophytes of vascular plants tend to be
   a smaller and to have smaller sex organs
   b smaller but to have larger sex organs
   c larger and to have larger sex organs
   d larger but to have smaller sex organs

4. Heterospory is the production of
   a sexual and asexual spores
   b large and small spores
   c haploid and diploid spores
   d diploid and tetraploid spores

5. A leaf that bears a sporangium is called a
   a sporidium
   b sporophore
   c sporangiophore
   d sporophyll

6. A microphyll is a leaf with
   a one unbranched vascular strand
   b one branched vascular strand
   c two unbranched, parallel vascular strands
   d no vascular tissue

7. A tapetum is
   a strengthening tissue in aerial stems
   b strengthening tissue in gametophytes
   c nutritive tissue in a sporangium
   d nutritive tissue in an archegonium

8. A strobilus is
   a the central tissue of an adventitious root
   b a group of sporophylls tightly clustered at the apex of a stem
   c a reproductive structure found in fossil members of the Psilopsida
   d a structure that assists in spore dissemination

9. Separate male and female gametophytes develop within the walls of the spores from which they arise in
   a *Rhynia*
   b *Lycopodium*
   c *Psilotum*
   d *Selaginella*

10. The gametophytes of *Psilotum* obtain their food from
    a photosynthesis
    b the sporophytes
    c mycorhizal fungi

11. *Rhynia* is a fossil plant that had only
    a leaves and stems
    b stems and sporangia
    c sporangia and roots
    d roots and leaves

1 ______
2 ______
3 ______
4 ______
5 ______
6 ______
7 ______
8 ______
9 ______
10 ______
11 ______

BASIC FACTS

The Tracheophyta are vascular plants; their sporophytes have true stems, leaves, and roots supplied with xylem and phloem. Sporophytes are relatively large, photosynthetic, and independent; gametophytes are small and may be either dependent on the sporophyte or independent. Tracheophytes are typically terrestrial plants; a cuticle reduces transpiration, and stomata permit gaseous exchange with the atmosphere. Some tracheophytes require water for fertilization. Sex organs are smaller than those of the Bryophyta.

The division TRACHEOPHYTA has four subdivisions: PSILOPSIDA (psilopsids), LYCOPSIDA (club mosses), SPHENOPSIDA (horsetails), and PTEROPSIDA (ferns and seed plants). The first three subdivisions have leaves called MICROPHYLLS, which are scalelike, have a single, unbranched vascular bundle, and are usually small. The gametophyte is called a PROTHALLUS.

The PSILOPSIDA have only two living genera, one of which is *Psilotum*, the simplest living vascular plant known. *Psilotum* has a rhizome from which arise upright, photosynthetic, aerial stems that are naked except for a few minute, scalelike leaves, some of which have three fused sporangia in their axils. Rhizomes and aerial stems exhibit dichotomous branching (branching into two equal or nearly equal branches). Gametophytes are small, dichotomously branched plants that lack chlorophyll; they grow in soil as separate plants and obtain their food from mycorhizal fungi. The Psilopsida have no roots, but the rhizomes and gametophytes have rhizoids.

The CLUB MOSSES (LYCOPSIDA) have aerial stems covered with many spirally arranged leaves. Rhizomes have adventitious roots. Sporangia are borne on, or in the axils of, leaves

(Continued on page 84)

ADDITIONAL INFORMATION

The stems and roots of the TRACHEOPHYTA have vascular tissue surrounded by a cortex; the outermost tissue is the epidermis. Apical meristems account for growth in length of stems and roots. Sporangia usually have a nutritive tissue, called a TAPETUM, that lies just inside the sporangium wall and nourishes the sporogenous tissue.

The rhizomes and aerial stems of *Psilotum* have a central core of xylem surrounded by phloem. The xylem contains tracheids, but no vessels. Leaves contain no vascular tissue. Sporangia crack open at maturity, releasing spores. The subterranean gametophytes lack chlorophyll but are independent of the sporophytes; they obtain their food from the soil through the activities of mycorhizal fungi. Spherical antheridia without stalks grow on the surfaces of the gametophytes. The venters of archegonia are embedded in the gametophytes; only their necks extend beyond the surface. Sperms are motile by means of many flagella.

The stems and roots of *Lycopodium* have a central core of vascular tissue. Xylem has tracheids but no vessels. Branching horizontal stems give rise to erect branches. Each leaf contains a single vascular strand connected by a LEAF TRACE to the vascular tissue of the stem. Sporangia crack open at maturity, releasing spores. Antheridia and the venters of archegonia are embedded in the upper surfaces of the gametophytes. Sperms are biflagellate.

Selaginella has horizontal stems from which may arise upright stems. Each leaf has a single leaf trace. The xylem contains tracheids and, in some species, vessels. Adventitious roots grow from special structures, called RHIZOPHORES, that arise from the stems. The lower sporophylls of a strobilus are usually megasporophylls, the upper ones microsporophylls. Sporangia break open at maturity; microspores are released shortly after they begin to germinate, but megaspores may not be shed. Gametophytes are very much reduced; the megagametophyte has some photosynthetic tissue and several archegonia, but the microgametophyte is nonphotosynthetic and consists of only one vegetative cell and an antheridium. Sperms are biflagellate. The retention of the megagametophyte in the megasporangium until a new embryo is formed resembles SEED FORMATION. A true seed, however, is surrounded by one or two integuments; there are no integuments around the megasporangium of *Selaginella*.

In most species of *Equisetum* the aerial stems are all green and bear a single strobilus at the tip, but other species have separate vegetative and fertile shoots, the latter being brown and nonphotosynthetic. Stems contain a ring of vascular bundles around a hollow pith. Epidermal cell walls are impregnated with silica compounds. Each leaf has a single

(Continued on page 84)

EXPLANATIONS

1. *All tracheophytes have* independent sporophytes; gametophytes are dependent or independent according to the species.

2. *Two features that the living Psilopsida, Lycopsida, and Sphenopsida have in common are* microphylls and lack of seeds, although a few living and fossil members possess structures having some characteristics of seeds. Vessels and secondary growth are extremely rare in these groups, although many fossil members had cambial activity.

3. *Compared with the gametophytes of the bryophytes, the gametophytes of vascular plants tend to be* smaller and to have smaller sex organs. One evolutionary trend among the vascular plants is the gradual reduction in size of gametophytes and their reproductive organs.

4. *Heterospory is the production of* large and small spores that after germination form female and male gametophytes, respectively.

5. *A leaf that bears a sporangium is called a* sporophyll. The small modified branch system in *Equisetum* that bears sporangia is called a sporangiophore.

6. *A microphyll is a leaf with* one unbranched vascular strand. There is no gap in the vascular bundle of the stem at the point of departure of the leaf trace. Microphylls are usually, but not necessarily, smaller than megaphylls.

7. *A tapetum is* nutritive tissue in a sporangium; it provides nourishment to the developing spores.

8. *A strobilus is* a group of sporophylls tightly clustered at the apex of a stem. Usually they are visibly different from ordinary vegetative leaves.

9. *Separate male and female gametophytes develop within the walls of the spores from which they arise in Selaginella. Selaginella* shows some of the steps in the assumption of the seed habit, although it does not have true seeds.

10. *Gametophytes of Psilotum obtain their food from* mycorhizal fungi. These gametophytes are subterranean and are completely or partially devoid of chlorophyll. Some species of *Lycopodium* also have subterranean gametophytes that depend on mycorhizal fungi.

11. *Rhynia is a fossil plant that had only* stems and sporangia. Underground stems and their rhizoids performed the functions of roots and root hairs as they do in *Psilotum. Rhynia* had no leaves; the aerial stems were photosynthetic.

Answers

c	1
a	2
a	3
b	4
d	5
a	6
c	7
b	8
d	9
c	10
b	11

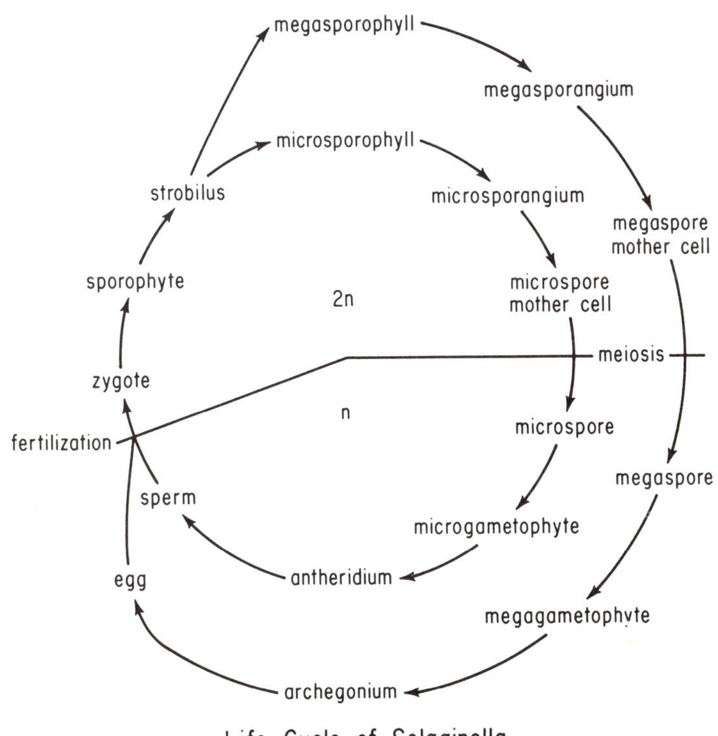

Life Cycle of Selaginella

called SPOROPHYLLS. In many species sporophylls are grouped together at the tips of stems, forming a CONE or STROBILUS. Spores of *Lycopodium* (ground pine) fall to the ground and germinate into small gametophytes that in some species grow on the soil surface and are green and photosynthetic, and in other species are subterranean and colorless saprophytes that obtain some of their food from a mycorhizal fungus. Each species of *Lycopodium* is HOMOSPOROUS, that is, it forms only one type of spore, and the gametophytes produced by the spores are all alike. In HETEROSPOROUS species, such as those of genus *Selaginella*, the plants produce two types of spores. After germination, the larger spores (MEGASPORES) form relatively large female gametophytes (MEGAGAMETOPHYTES), and the smaller spores (MICROSPORES) form small male gametophytes (MICROGAMETOPHYTES). Megaspores and microspores are usually produced in separate sporangia called MEGASPORANGIA and MICROSPORANGIA, respectively, and the sporophylls on which the sporangia are borne are called MEGASPOROPHYLLS and MICROSPOROPHYLLS, respectively. Gametophytes are very small and develop within the spore walls, before or after spores are shed. Occasionally fertilization and even early development of a young sporophyte occur before the megaspore is shed. Usually only one sporophyte is produced by a megagametophyte.

Equisetum is the only living HORSETAIL (SPHENOPSIDA). Jointed, hollow aerial stems arise from rhizomes with adventitious roots. Narrow leaves are arranged in whorls at nodes. A single STROBILUS occurs at the tip of a stem. Branching off from its main axis are SPORANGIOPHORES, each of which bears from five to ten sporangia. All species are HOMOSPOROUS; each spore is equipped with four spoon-shaped ELATERS that assist in spore dissemination. The gametophyte is a small, flattened, independent plant with an irregularly lobed upper surface. The lobes are photosynthetic.

vascular trace. Gametophytes are small, independent, photosynthetic plants. Archegonia and antheridia are similar to those of *Lycopodium*. Usually archegonia are produced first; then if fertilization does not occur antheridia are formed. Sperms bear many flagella.

Rhynia, the simplest vascular plant known to have existed, is a FOSSIL member of the PSILOPSIDA. It consisted of a dichotomously branched rhizome with dichotomously branched aerial stems. The plants had no roots or leaves. Fertile branches terminated in single sporangia. Other fossil members of the Psilopsida had leaves, but all were rootless. It has been hypothesized that a species similar to *Rhynia* was the ancestor of vascular plants. No gametophytes of *Rhynia* have been discovered, but since both gametophytes and sporophytes of *Psilotum* are cylindrical, dichotomously branched structures, it is further hypothesized that the two generations of the ancestral plant were similar to each other in form if not in size. This ancestor may have evolved from a green alga with isomorphic alternation of generations.

Modern members of the LYCOPSIDA and SPHENOPSIDA usually do not exceed two or three feet in height, but some FOSSIL members were large trees reaching heights of up to 100 feet and possessing secondary tissues produced by vascular cambium and cork cambium. *Lepidodendron* and *Sigillaria* were heterosporous tree lycopods with tall straight trunks. *Lepidodendron* had relatively small leaves spirally arranged on the stems, which terminated in strobili. The long, strap-shaped leaves of *Sigillaria* were borne in dense tufts at the tips of branches; strobili grew below the tufts of leaves. Some lycopods produced structures that resembled seeds but lacked integuments. These plants are not believed to be ancestral to modern seed plants. Forests of tree lycopods grew in swamps filled with stagnant, anaerobic water. When the trees died and fell into the water, decomposition proceeded very slowly. The tremendous pressure of many layers of sediment that later formed above them compressed the partially decomposed material into COAL. A relatively small degree of compression produced LIGNITE, a poor grade of coal; more compression produced BITUMINOUS, or soft, coal; and still more compression produced ANTHRACITE, or hard coal. Most coal was formed from lycopods, ferns, and some early gymnosperms. *Calamites* was a tall fossil member of the Sphenopsida; its branches and its narrow leaves were borne in whorls. Some species were heterosporous.

22 FERNS

SELF-TEST

1. Ferns differ from the other members of the Pteropsida in
   a having leaf gaps
   b lacking seeds
   c being heterosporous
   d lacking vascular tissue

2. Fern gametophytes are typically
   a small, green, and homothallic
   b small, nongreen, and heterothallic
   c large, green, and heterothallic
   d large, nongreen, and homothallic

3. Which of the following is *not* involved in the fertilization of ferns?
   a pollen tube
   b water
   c archegonium
   d flagellated sperm

4. Sporangia of most ferns are found
   a in strobili
   b on rhizomes
   c on the upper surface of the leaf
   d on the lower surface of the leaf

5. A cluster of closely associated sporangia is called a(n)
   a indusium
   b inflorescence
   c sorus
   d sporangiophore

6. Spore dissemination in many ferns is effected by the
   a annulus
   b indusium
   c tapetum
   d sorus

7. The developing fern embryo obtains food by means of
   a a green primary leaf
   b a green primary stem
   c the foot
   d the root

8. In most ferns apical growth occurs in
   a stems
   b stems and roots
   c stems, roots, and leaves
   d stems, roots, leaves, and gametophytes

9. The formation of a sporophyte from a vegetative portion of a prothallus is called
   a apocarpy
   b apogamy
   c apomixis
   d apospory

10. A stele without a central pith is called a
   a dictyostele
   b protostele
   c siphonostele
   d solenostele

1 ______
2 ______
3 ______
4 ______
5 ______
6 ______
7 ______
8 ______
9 ______
10 ______

BASIC FACTS

The subdivision PTEROPSIDA includes those vascular plants that have relatively large leaves with one or more branching vascular bundles; LEAF GAPS, formed by the departure of the leaf trace from the vascular tissue of the stem, are characteristic of this group. The Pteropsida is divided into three classes: FILICINEAE (ferns), GYMNOSPERMAE (plants with naked seeds), and ANGIOSPERMAE (flowering plants).

Most FERNS grow in moist areas of the tropical and temperate zones, but some species live in deserts and others in Arctic regions. Both generations of a typical fern are green and independent, but the sporophyte is the conspicuous one. The SPOROPHYTE usually consists of a branched rhizome with its adventitious roots and large, compound leaves called FRONDS. Each leaflet is called a PINNA; in doubly compound leaves each division of a pinna is called a PINNULE. A FIDDLEHEAD is a young leaf with its blade still tightly coiled. SPORANGIA occur in clusters called SORI on the lower surface of the leaf. A sorus may or may not be covered by a tissue called the INDUSIUM. In some species in which the sporangia occur near the margin, the margin may curl over them and form a FALSE INDUSIUM. Several spore mother cells in the sporangium undergo meiosis and form haploid spores. The sporangial wall has a ring of tissue called an ANNULUS which, by drying and contraction, forcibly discharges the spores. A spore germinates into a short green filament, the apical portion of which forms the adult prothallus. Most ferns produce sporangia on ordinary vegetative leaves, but a few species are DIMORPHIC; their sporangia are formed on separate leaves called SPOROPHYLLS or on only a few pinnae of a leaf. The fertile leaves of dimorphic species may not be photosynthetic and may appear morphologically

(Continued on page 88)

ADDITIONAL INFORMATION

SPORE DISCHARGE in the ferns is effected by the annulus. The ANNULUS is a single ring of cells that partially encircles the sporangium; it is interrupted by a few cells called LIP CELLS that have thin walls. The cells of the annulus have thick walls on three sides: the inner side and the two sides that touch adjacent annulus cells. Other walls of these cells are thin, and through them water evaporates as the sporangium grows older. As the water content decreases, the cohesion of water molecules to each other and the adhesion of water molecules to the thick cell walls pull these walls together. The annulus cells shrink, the annulus shortens and straightens, and a tear begins to form at the weaker lip cells and continues across the base of the sporangium. The top of the sporangium, which has most of the spores in it, bends backward. As more water evaporates from the annulus cells, the tension on the remaining water becomes so great that the water is suddenly converted to water vapor. Since there is no longer any tension exerted by water on the cell walls, they quickly return to their original position. The annulus curves again, and the top of the sporangium snaps back in place, hurling its spores out like stones from a slingshot. This mechanism, which is found only among ferns, insures that spores will be disseminated in dry weather when they have the greatest chance of being carried a long distance.

A STELE is the central cylinder of a stem or root. There are two main types: a PROTOSTELE, which is a solid core of primary vascular tissue, and a SIPHONOSTELE, which is a ring of primary vascular tissue surrounding a central column of pith. In a protostele, xylem usually occupies the center of the stele and is surrounded by phloem, but the two tissues may intermingle. The stems of the Psilopsida, the Lycopsida, and some primitive ferns have protosteles. The ring of a siphonostele usually has one or more leaf gaps at any one level. A siphonostele with only one leaf gap appearing in a cross section is called a SOLENOSTELE; some fern rhizomes have solenosteles. If several leaf gaps appear in the cross section of a siphonostele, it is a DICTYOSTELE. Dictyosteles are found in stems of *Equisetum*, many ferns, and seed plants. Protosteles are the most primitive type of stele, dictyosteles the most advanced.

The XYLEM of ferns contains tracheids, but vessels are very rare in this group. There is no cambium, and no secondary growth occurs. Most ferns do not have aerial stems and do not grow very tall, but a few tropical species are trees.

In many ferns, ROOT and STEM cells derive from one apical cell rather than from a group of apical cells (apical meristem) as they do in the seed plants. Fern LEAVES are unusual in having an apical meristem that persists for a long time; in a few species leaves may grow to a length of 100 feet. The apical cell of the gametophyte lies in the notch; daughter cells

(Continued on page 88)

EXPLANATIONS

1. *Ferns differ from the other members of the Pteropsida in* lacking seeds. All members have leaf gaps and vascular tissue.

2. *Fern gametophytes are typically* small, green, and homothallic. In only a few species gametophytes are subterranean and lack chlorophyll and live symbiotically with fungi.

3. A pollen tube is not involved in the fertilization of ferns. Flagellated sperms swim through water to the archegonia where fertilization occurs.

4. *Sporangia of most ferns are found* on the lower surface of the leaf. In some species separate vegetative and reproductive leaves are produced.

5. *A cluster of closely associated sporangia is called a* sorus. Some sori are covered by indusia.

6. *Spore dissemination in many ferns is effected by the* annulus, which brings about the forcible discharge of spores.

7. *The developing fern embryo obtains food by means of* the foot, which is embedded in the venter. Later, when the primary leaf expands, the young sporophyte manufactures its own food.

8. *In most ferns apical growth occurs in* stems, roots, leaves, and gametophytes. Usually each organ has a single apical cell from which the other cells derive.

9. *The formation of a sporophyte from a vegetative portion of a prothallus is called* apogamy because gametes are not involved.

10. *A stele without a central pith is called a* protostele. This is the most primitive type of stele.

Answers

b	1
a	2
a	3
d	4
c	5
a	6
c	7
d	8
b	9
b	10

Characteristics of Nonseedforming Vascular Plants

	Psilotum	*Lycopodium*	*Selaginella*	*Equisetum*	*Ferns*
SPOROPHYTES					
Stems	Rhizomes and aerial stems	Rhizomes and aerial stems	Rhizomes and aerial stems	Rhizomes and aerial stems	Rhizomes (aerial stems in tree ferns)
Roots	Absent	Present	Present	Present	Present
Leaves	Microphylls	Microphylls	Microphylls	Microphylls	Megaphylls
Sporangia	Homosporous, in axils of leaves	Homosporous, on sporophylls in strobili	Heterosporous, on sporophylls in strobili	Homosporous, on sporangiophores in strobili	Homosporous, on variously modified sporophylls
GAMETOPHYTES					
Nutrition	Mycorhizal fungi	Photosynthesis or mycorhizal fungi	Parent sporophyte, limited photosynthesis	Photosynthesis	Photosynthesis (mycorhizal fungi in a few cases)
Sperms	Multiflagellate	Biflagellate	Biflagellate	Multiflagellate	Multiflagellate

different from the vegetative leaves. Most ferns are HOMOSPOROUS, but a few aquatic ferns are HETEROSPOROUS. Some fossil ferns were heterosporous, but the so-called seed ferns are now classified as gymnosperms.

The GAMETOPHYTE is a small, inconspicuous, flat, heart-shaped PROTHALLUS that grows on the surface of the ground. Gametophytes are usually bisexual (homothallic) with archegonia, antheridia, and rhizoids developing on the ventral surface. Archegonia occur near the notch. Neither antheridia nor archegonia have stalks. The venter of the archegonium is embedded in the prothallus, and the neck, a few cells long, extends beyond the surface. Each archegonium has a single egg. Many multiflagellate sperms are released from the antheridia; they require water in which to swim toward the archegonia. A zygote develops into an embryo divided into four parts: FOOT, STEM, PRIMARY ROOT, and PRIMARY LEAF; there is no suspensor. The foot remains in the venter and absorbs food from it. The primary root grows downward into the soil, and the primary leaf curves upward around the gametophyte and soon produces food. After the young sporophyte becomes established, the gametophyte dies. The stem is a rhizome from which adventitious roots and additional leaves grow; the primary root dies early.

ASEXUAL REPRODUCTION may occur by the dying back of old portions of the rhizomes; this separates the branches into separate plants.

cut off on two sides contribute to the formation of either side of the heart-shaped gametophyte.

In the usual life cycle of a member of the Embryophyta, a diploid sporophyte results from the union of an egg and a sperm, and a haploid gametophyte arises from a spore, the haploid product of meiosis. The ALTERNATION OF GENERATIONS is a part of the normal sexual cycle. In some ferns and mosses, sporophytes may be produced VEGETATIVELY from gametophytes, and gametophytes may be produced vegetatively from sporophytes. Vegetative cells of the prothallus, usually those near the archegonium, may organize into an embryo that develops into a mature, haploid sporophyte. Since gametes are not involved, this process is called APOGAMY. Vegetative cells of a sporophyte, especially cells in young leaves, may produce diploid gametophytes. Since spores are not involved, this process is called APOSPORY. A few species of ferns regularly exhibit apogamy or apospory. Apogamy and apospory demonstrate that chromosome numbers alone do not determine the characteristics of gametophyte and sporophyte generations.

The subdivisions Psilopsida, Lycopsida, Sphenopsida, and Pteropsida are thought to represent four separate EVOLUTIONARY LINES that developed independently—possibly from different groups of green algae—although some botanists think that the earliest Psilopsida may have been the ancestors of the other three. It is likely that green algae were the ultimate ancestors of all vascular plants.

23 GYMNOSPERMS

SELF-TEST

1. All seed plants are
   a homosporous and have dependent gametophytes
   b homosporous and have independent gametophytes
   c heterosporous and have dependent gametophytes
   d heterosporous and have independent gametophytes

2. Seed plants are unique among the Embryophyta in
   a being independent of external water for fertilization
   b being heterothallic and heterosporous
   c having alternation of generations
   d having some members that are trees

3. A seed is composed of tissue belonging to
   a two sporophyte generations
   b two gametophyte generations
   c two sporophyte generations and one gametophyte generation
   d two gametophyte generations and one sporophyte generation

4. Most gymnosperms differ from most angiosperms in having
   a seeds borne naked on scales
   b seeds covered by pericarps
   c wood with vessels
   d phloem with sieve tubes

5. Most gymnosperms have
   a both archegonia and antheridia
   b archegonia but no antheridia
   c antheridia but no archegonia
   d no antheridia or archegonia

6. The ancestors of the conifers are thought to be
   a ginkgoes
   b cycads
   c Pteridospermae
   d Cordaitales

7. Cycads differ from conifers in having
   a pinnately compound leaves and a great deal of secondary growth
   b pinnately compound leaves and little secondary growth
   c simple leaves and a great deal of secondary growth
   d simple leaves and little secondary growth

8. Cycads and conifers have
   a nonmotile sperms
   b motile sperms
   c nonmotile and motile sperms, respectively
   d motile and nonmotile sperms, respectively

9. In gymnosperms a pollen chamber is
   a the microsporangium in which pollen grains develop
   b the cell in the pollen grain in which the sperms are formed
   c a cavity in the ovule in which pollen grains are stored after pollination
   d an opening in the megagametophyte through which the pollen tube approaches the egg

10. A pollen grain is a
   a megaspore
   b microspore
   c megagametophyte
   d microgametophyte

1 ______
2 ______
3 ______
4 ______
5 ______
6 ______
7 ______
8 ______
9 ______
10 ______

BASIC FACTS

The classes GYMNOSPERMAE and ANGIOSPERMAE comprise the SEED PLANTS. All seed plants are heterosporous. Sporophytes of seed plants are green and independent; gametophytes lack chlorophyll and are completely dependent on the sporophytes. In general the sporophytes are larger than those of the ferns, and the gametophytes are smaller and simpler. Unlike the ferns, seed plants do not require water for fertilization; pollen grains are carried to the ovules by wind and reach the archegonia by means of pollen tubes. Seeds of gymnosperms are borne exposed on the surface of cone scales rather than being enclosed in a pericarp as they are in the angiosperms.

Two main groups of living gymnosperms are the cycads (order Cycadales) and the conifers (order Coniferales). The CYCADS are a small group of slow-growing trees of tropical and subtropical regions. Stems are usually unbranched and have a great deal of cortex and pith but only a small amount of vascular tissue. Some secondary growth occurs, but it is very slow. Leaves are large, fernlike, and pinnately compound. Cycads are dioecious, an individual plant bearing either staminate (male) or ovulate (female) cones. Most CONIFERS are evergreens of temperate and subarctic regions of the world. Leaves are usually needle-shaped or scalelike. Conifers include pine, hemlock, spruce, and redwood; most of them are trees with a great deal of secondary growth. Tracheids and ray cells are common in the xylem, but vessels are not present. Phloem has sieve cells, but no sieve tubes or companion cells. Most conifers are monoecious; they have staminate and ovulate cones on the same plant.

The OVULATE CONES of the cycads have megasporophylls that are usually leaflike and that bear two or more ovules on their margins. Ovulate cones of conifers are woody and con-

(Continued on page 92)

ADDITIONAL INFORMATION

The GAMETOPHYTES of gymnosperms are much smaller and simpler than those of most ferns. A young MEGAGAMETOPHYTE consists of a single cell with many free nuclei; later, cell walls form, making it multicellular. The archegonia are very much reduced, consisting of only one large egg cell, a short neck, and a ventral canal cell which dies early. A megagametophyte has only a few archegonia, often only two. The megagametophyte of the gymnosperms is sometimes called the ENDOSPERM; it is of different origin than the polyploid endosperm of flowering plants.

The mature MICROGAMETOPHYTE consists of only a few cells and has no antheridia. The POLLEN GRAIN usually has a GENERATIVE CELL, a TUBE CELL, and one or two PROTHALLIAL CELLS when it is shed. While the pollen grain is retained in the pollination chamber of the ovule it absorbs food from the nucellus. The generative cell divides into a STALK CELL and a BODY CELL; the latter divides into two SPERMS that enter the POLLEN TUBE produced by the tube cell. The pollen tube grows through the nucellus toward the archegonia and discharges the sperms, one of which fertilizes an egg; the other sperm and the remainder of the microgametophyte disintegrate. Pollen grains of gymnosperms are disseminated almost entirely by wind; fertilization is thus independent of an external supply of water, even in those species that have motile sperms.

Usually only one archegonium in a megagametophyte produces an embryo. In conifers, the zygote nucleus undergoes a series of divisions followed by cell-wall formation. This PROEMBRYO differentiates into four embryos and into SUSPENSOR cells, which push the embryos into the megagametophyte. Only one embryo lives. The ovule ripens into a SEED consisting of three generations: the adult sporophyte, represented by the integument and the nucellus; the megagametophyte; and the young embryo sporophyte. When mature, the embryo has an EPICOTYL (or PLUMULE), a HYPOCOTYL, a RADICLE, and two or more COTYLEDONS. In some conifers, such as the pines, formation of an ovulate cone and its seeds often requires two years, pollination occurring in the first year and fertilization in the second. In many conifers the seeds are equipped with wings and are wind-disseminated. Cycad seeds and those of some conifers germinate soon after they are shed; those of other conifers require a period of dormancy. Germination is similar to that of flowering plants. The radicle is the first organ to penetrate the seed coat.

Stems and roots of seed plants grow in length by the activity of APICAL MERISTEMS that have many dividing cells. In those species that have secondary growth, CAMBIUM is responsible for increase in width of these organs. The wood of many conifers contains RESIN produced by resin ducts.

(Continued on page 92)

EXPLANATIONS

1. *All seed plants are* heterosporous and have dependent gametophytes. Both gymnosperms and angiosperms produce small, nonphotosynthetic microgametophytes and megagametophytes.

2. *Seed plants are unique among the Embryophyta in* being independent of external water for fertilization. Sperms are produced after pollen grains reach the ovules.

3. *A seed is composed of tissue belonging to* two sporophyte generations and one gametophyte generation. The young embryo sporophyte develops within the megagametophyte which is covered by tissue belonging to the parent sporophyte.

4. *Most gymnosperms differ from most angiosperms in having* seeds borne naked on the scales of ovulate cones. Scales of young cones are so tightly packed together that the young seeds are enclosed in the cones; at maturity the scales spread apart, exposing the seeds. In angiosperms the seeds are enclosed in pericarps. Most gymnosperms have no vessels in the xylem or sieve tubes in the phloem.

5. *Most gymnosperms have* archegonia but no antheridia. Both megagametophytes and microgametophytes are very small, but in most species a few archegonia are present in the megagametophytes. Very few species lack both antheridia and archegonia.

6. *The ancestors of the conifers are thought to be* Cordaitales, which were tall cone-bearing trees.

7. *Cycads differ from conifers in having* pinnately compound leaves and little secondary growth. Vegetatively, cycads resemble ferns more than they do conifers, but cycads are classified with conifers in the Gymnospermae because the reproductive structures and life cycles of these two groups are similar.

8. *Cycads and conifers have* motile and nonmotile sperms, respectively. Cycad sperms are multiflagellate cells; those of most conifers are nonmotile like those of flowering plants.

9. *In gymnosperms a pollen chamber is* a cavity in the ovule in which pollen grains are stored after pollination. Pollen often is shed many months before the megagametophytes and their archegonia are ready for fertilization. The pollen chamber of the gymnosperms should not be confused with the pollen chambers in the anthers of the stamens in angiosperms.

10. *A pollen grain is a* microgametophyte that has germinated from a microspore and that remains within the microspore wall.

Answers

c	1
a	2
c	3
a	4
b	5
d	6
b	7
d	8
c	9
d	10

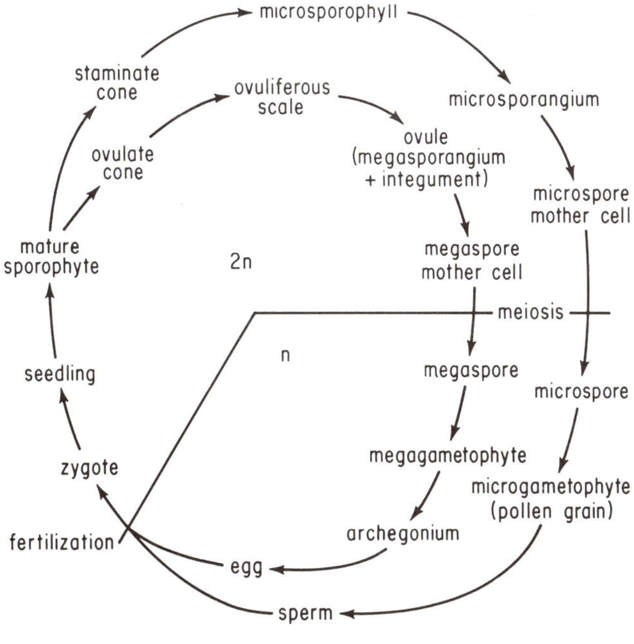

Life Cycle of a Conifer

sist of a series of bracts, each with an OVULIFEROUS SCALE; the scale is more complex than a single modified leaf and is not called a megasporophyll. Two or more ovules are borne on the upper surface of an ovuliferous scale. A gymnosperm OVULE consists of a MEGASPORANGIUM surrounded by an INTEGUMENT with an opening called the MICROPYLE. The megasporangium wall is the NUCELLUS. Between the nucellus and the integument is a POLLEN CHAMBER (MICROPYLAR CHAMBER). Within the megasporangium there is one megaspore mother cell which by meiosis produces four megaspores, three of which disintegrate. The remaining megaspore germinates within the megasporangium, and there it produces a small megagametophyte. A few archegonia are embedded in the megagametophyte.

The microsporophylls of the STAMINATE CONES bear MICROSPORANGIA (POLLEN SACS) on their lower surfaces. The microsporangia contain many microspore mother cells which undergo meiosis, each producing four microspores. Microspores germinate within the microsporangium and form small microgametophytes, or POLLEN GRAINS. Pollen grains are disseminated by wind. Any that land on an ovule near the micropyle are drawn into the pollen chamber where they may live for several weeks or months. During this time sperms are produced and a POLLEN TUBE grows toward an archegonium. Fertilization occurs after discharge of the sperms from the pollen tube. The sperms of cycads are motile by means of many flagella; those of conifers are nonmotile sperm nuclei which are discharged directly into the archegonium. Usually only one embryo matures in an ovule. The megagametophyte is the main food source of the embryo. The integument matures into a SEED COAT, and the ovule becomes a SEED.

Ginkgo biloba, the maidenhair tree, is the only living species of the order GINKGOALES. It is often classified with the conifers, for like the conifers it is a tall tree with a great deal of wood. However, like the cycads, *Ginkgo* is DIOECIOUS and has motile sperms. The cones are so modified that they bear little external resemblance to typical gymnosperm cones. The microsporophylls look very much like the stamens of flowering plants. The ovulate cone bears only two ovules; usually one of them matures into a fleshy seed that superficially resembles a fruit. *Ginkgo* is one of the few DECIDUOUS gymnosperms, its flat, fan-shaped leaves dropping in autumn.

Cycads and conifers represent two different EVOLUTIONARY LINES. SEED FERNS (Order PTERIDOSPERMALES), now extinct, are considered to be the ancestors of modern CYCADS. In their vegetative features cycads resemble ferns; they have compound leaves and stems with a great deal of cortical tissue but little wood. Seed ferns had no cones; however, they did have true seeds, usually borne on the margins of vegetative leaves, and pollen grains. Many fossil seed ferns had well-developed secondary wood.

The BENNETTITALES, also extinct, were probably descendants of the seed ferns. They were only a few feet high; most species had thick, unbranched stems that bore a crown of pinnately compound leaves and short axillary fertile shoots. In some species the pinnately compound microsporophylls were found beneath the female cones; other species were dioecious.

The origin of CONIFERS is uncertain. An extinct group of gymnosperm trees, the CORDAITALES, had wood that resembled that of modern conifers. They had tall trunks, branched at the top, and long, strap-shaped leaves. The small unisexual cones bore sterile appendages among the fertile ones. The Cordaitales are one group considered by some botanists to be ancestors of the conifers.

24 ANGIOSPERMS

SELF-TEST

1. Three features that are typical of angiosperms but are rare or unknown in other plants are
   a flowers, seeds, and tracheids
   b flowers, fruits, and vessels
   c fruits, tracheids, and nonmotile sperms
   d seeds, vessels, and nonmotile sperms

2. A stamen is a
   a microsporangium
   b microsporophyll
   c megasporangium
   d megasporophyll

3. Angiosperms differ from gymnosperms in having
   a tracheids
   b antheridia
   c vessels
   d archegonia

4. The gametophytes of angiosperms have
   a antheridia and archegonia
   b antheridia but no archegonia
   c archegonia but no antheridia
   d neither archegonia nor antheridia

5. The endosperm of angiosperms is
   a the megagametophyte
   b the microgametophyte
   c part of the embryo
   d a triploid tissue

6. In a hypogynous flower the sepals, petals, and stamens arise
   a below the ovary
   b above the ovary
   c on the hypanthium which forms a cup around the ovary
   d on the hypanthium which is fused to the ovary

7. In an ovary with parietal placentation the ovules arise from the
   a dividing walls in the ovary
   b ovary wall
   c central axis of the ovary
   d base of the ovary

8. The dicots and monocots are
   a primitive and advanced groups, respectively
   b advanced and primitive groups, respectively
   c two separate evolutionary lines
   d two artificial groups that have no evolutionary relationship to each other

9. A combination of flower parts that would probably indicate a monocot is
   a 3 sepals, 3 petals, 6 stamens, and 3 carpels
   b 4 sepals, 4 petals, 4 stamens, and 4 carpels
   c 5 sepals, 5 petals, 5 stamens, and 5 carpels
   d 5 sepals, 5 petals, 10 stamens, and 5 carpels

10. A highly evolved angiosperm would be likely to have
    a hypogynous flowers with a few fused parts arranged in a spiral
    b hypogynous flowers with many separate parts arranged in whorls
    c epigynous flowers with a few fused parts arranged in whorls
    d epigynous flowers with many separate parts arranged in a spiral

11. The maturation at different times of stamens and pistils in the same flower is called
    a heterospory
    b heterostyly
    c dichogamy
    d dichotomy

1 ________
2 ________
3 ________
4 ________
5 ________
6 ________
7 ________
8 ________
9 ________
10 ________
11 ________

BASIC FACTS

All plants producing flowers and fruits belong to the class ANGIOSPERMAE. Unlike the gymnosperms, angiosperms have ovules enclosed within carpels, and their seeds are borne within fruits. Sporophytes may be trees or shrubs in both groups, but many angiosperms are herbaceous. Gymnosperms are all perennials; angiosperms are either perennials, biennials, or annuals. Vascular tissue of angiosperms is more complex than that of gymnosperms, the former possessing vessels and companion cells which the latter lack. Most gymnosperms are evergreen, whereas many angiosperms are deciduous. Cones of gymnosperms are unisexual; flowers of primitive angiosperms are bisexual, but those of some advanced species are unisexual. The gametophytes of angiosperms are more reduced than those of gymnosperms. The mature megagametophyte consists of a seven-celled embryo sac, the mature microgametophyte of only three cells, one being the tube cell and the other two being sperms. No members of the Angiospermae possess motile sperms. Neither antheridia nor archegonia are present in flowering plants. Gymnosperm pollen grains land directly on the ovule and germinate within the pollen chamber, whereas pollen grains of angiosperms germinate on the stigma, the pollen tube growing down through the style to the ovule.

FLOWERS are short stems bearing four types of modified leaves: SEPALS, PETALS, MICROSPOROPHYLLS (STAMENS), and MEGASPOROPHYLLS (CARPELS). The POLLEN SACS are MICROSPORANGIA; within them the MICROSPORE MOTHER CELLS (pollen mother cells) produce MICROSPORES by meiosis. The young MICROGAMETOPHYTES (POLLEN GRAINS) form within the microspore walls.

Each OVULE has a single MEGASPORANGIUM; the sporangial wall is the NUCELLUS. The megasporangium con-

(Continued on page 96)

ADDITIONAL INFORMATION

The angiosperms are a large, diverse group. The dicots and monocots represent two different lines of evolution within the angiosperms; the monocots probably arose from some primitive dicots. Within each of these two groups are plants that have flowers with mostly primitive characteristics, others with mostly advanced characteristics, and still others that are intermediate, having various mixtures of primitive and advanced characteristics. PRIMITIVE FLOWERS have MANY FLOWER PARTS SPIRALLY ARRANGED on the receptacle. ADVANCED FLOWERS have FEW FLOWER PARTS, usually a definite number, arranged in WHORLS around the receptacle; some parts may be FUSED. Carpels are often fused into one pistil; sepals, petals, or filaments of stamens may be fused into a tube, or unlike parts may be fused to each other. Primitive flowers are COMPLETE (possessing all four flower parts) and are PERFECT (possessing both stamens and pistils). Advanced flowers are INCOMPLETE (lacking one or more flower parts) or IMPERFECT (lacking either stamens or pistils). ACTINOMORPHIC (radially symmetrical) or REGULAR flowers are more primitive than ZYGOMORPHIC (bilaterally symmetrical) or IRREGULAR flowers. HYPOGYNOUS flowers are those in which the sepals, petals, and stamens arise from the receptacle below the ovary; the ovaries of these flowers are SUPERIOR. In EPIGYNOUS flowers the sepals, petals, and stamens appear to arise from the top of the ovary; actually, they arise from a tissue called the HYPANTHIUM that is fused to the ovary wall. The hypanthium, sometimes called the CALYX or FLORAL TUBE, is considered to be the fused bases of sepals, petals, and stamens. The ovaries of epigynous flowers are INFERIOR. A few flowers have a hypanthium that is not fused to the ovary but forms a cup around it, the sepals, petals, and stamens appearing to arise from its rim; such flowers are PERIGYNOUS, and the ovary is SUPERIOR. Epigyny is a more advanced condition than hypogyny, and perigyny is an intermediate condition. In the most primitive angiosperms the flowers tend to be borne SINGLY, while they are often in INFLORESCENCES in more advanced species. Few angiosperms have only primitive features or only advanced features. The magnolia family is one of the most primitive families of angiosperms. The composite family is the most highly evolved of the dicots, as is the orchid family in the monocots.

The PLACENTA is the tissue in the ovary to which the ovules are attached. If the carpels of a compound pistil are fused so that their placentas meet in the center of the ovary and form an axis, and if the walls separate the cavities of the carpels, the placentation is AXILE; but if there are no separating walls and the fused placentas form a central stalk rising from the base of the ovary, the placentation is FREE CENTRAL. In a few species each carpel is not folded, but the margins of adjacent carpels are fused to each other around a central cavity; the wall of the ovary has several placentas, and the

(Continued on page 96)

EXPLANATIONS

1. *Three features that are typical of angiosperms but are rare or unknown in other plants are* flowers, fruits, and vessels. No other plants have flowers or fruits, and only a few other vascular plants have vessels. Gymnosperms have seeds and tracheids. Nonmotile sperms are found among the conifers as well as among some thallophytes.

2. *A stamen is a* microsporophyll bearing microsporangia.

3. *Angiosperms differ from gymnosperms in having* vessels in the xylem. Both angiosperms and gymnosperms have tracheids.

4. *The gametophytes of angiosperms have* neither archegonia nor antheridia. Angiosperm gametophytes are the simplest and most reduced gametophytes in the Embryophyta.

5. *The endosperm of angiosperms is* a triploid tissue that is found in no other plants. Double fertilization (one fertilization producing a zygote, the other the endosperm) is also unique to the angiosperms.

6. *In a hypogynous flower the sepals, petals, and stamens arise* below the ovary, directly from the receptacle.

7. *In an ovary with parietal placentation the ovules arise from the* ovary wall. Ovaries with parietal placentation have no dividing walls.

8. *The dicots and monocots are* two separate evolutionary lines within the angiosperms. Each group has its own primitive and advanced members. However, the monocots may have arisen from some primitive dicots.

9. *A combination of flower parts that would probably indicate a monocot is* 3 sepals, 3 petals, 6 stamens, and 3 carpels. Most monocot flowers have flower parts in multiples of three.

10. *A highly evolved angiosperm would be likely to have* epigynous flowers with a few fused parts arranged in whorls. How advanced or how primitive a plant is, is determined by a combination of characteristics rather than by just one.

11. *The maturation at different times of stamens and pistils in the same flower is called* dichogamy. Dichogamy is one mechanism that prevents self-pollination.

Answers

b	1
b	2
c	3
d	4
d	5
a	6
b	7
c	8
a	9
c	10
c	11

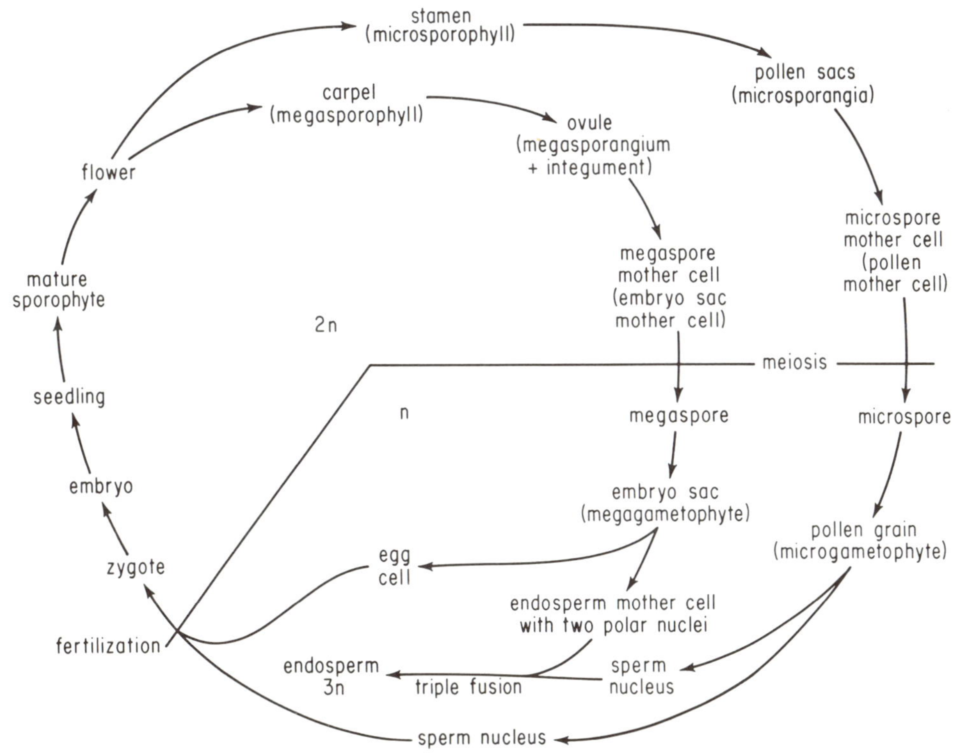

Life Cycle of an Angiosperm

tains only one MEGASPORE MOTHER CELL (embryo sac mother cell) which produces FOUR MEGASPORES by meiosis. One of these megaspores germinates into a multinucleate MEGAGAMETOPHYTE (EMBRYO SAC) that becomes multicellular by the formation of cell walls around the nuclei. One of these cells, the ENDOSPERM MOTHER CELL, which has two polar nuclei, is fertilized by a sperm nucleus. This fertilization, called TRIPLE FUSION, produces the ENDOSPERM. The EMBRYO is formed by the fusion of another sperm nucleus with the egg. The two fertilizations together are called DOUBLE FERTILIZATION, which is unique to flowering plants.

Unlike embryos of other members of the Embryophyta, angiosperm embryos are not nourished by the megagametophytes. The food supply of the embryo is stored either in the endosperm or in the cotyledons of the embryo. In the latter case, no endosperm is present in the mature seed. The megagametophyte disintegrates before the seed reaches maturity.

The angiosperms are divided into two subclasses: Monocotyledonae (MONOCOTS) and Dicotyledonae (DICOTS). Embryos of the former have one cotyledon; those of the latter have two. Flower parts of monocots are usually in groups of three or in multiples of three, while those of dicots occur in groups of five or four, or multiples thereof. Parallel venation is typical of monocot leaves, net venation of dicot leaves. Monocots generally have no secondary growth and no cambium; primary vascular bundles are scattered throughout the stem. Many dicots have secondary growth produced by vascular and cork cambiums; primary vascular bundles in stems usually are arranged in a single ring.

placentation is PARIETAL. In a SIMPLE PISTIL with one placenta running the length of the ovary the placentation is called either PARIETAL or MARGINAL. If only one ovule arises from the base of a simple ovary, the placentation is BASAL.

A few angiosperms are regularly SELF-POLLINATED and produce viable seed, and a few require CROSS-POLLINATION to set seed. Most species can reproduce either way, but they set more vigorous seed when cross-pollinated. A number of mechanisms either insure or encourage cross-pollination. In dioecious species (those with male and female flowers on separate plants) any pollination is a cross-pollination. In some other plants the stamens and pistils of the same flower mature at different times (DICHOGAMY). In SELF-INCOMPATIBLE species pollen will not germinate on stigmas of the same plant, or if it does, the pollen tube grows so slowly that the egg is no longer receptive when the tube reaches it. Some insect-pollinated species show HETEROSTYLY: some plants have flowers with long styles and short stamens, and others have short styles and long stamens. The evolution of many species of insects and that of many species of flowering plants were interrelated. Early angiosperms were wind-pollinated, but a mutual dependence evolved between the later, insect-pollinated angiosperms and their pollinating insects.

Angiosperms are the dominant plants in most of the land areas of the world today. Their habitats range from the tropics to arctic regions and from deserts to swamps and lakes. They are assumed to have originated from tropical vascular plants, possibly primitive gymnosperms, but their ancestors have not been determined with certainty. The evolution of the many species of mammals and birds seems to have depended on the evolution of the angiosperms. The great civilizations of the ancient world grew up in areas suitable for the cultivation of rice, wheat, corn, and other members of the grass family which still provide the major caloric intake of humans and most of their domestic animals. Members of the legume family (such as peas, beans, and alfalfa) are the only major agricultural plants with a high protein content. Angiosperms are also the sources of lumber (hardwoods), fibers (cotton, flax, ramie, jute), beverages (coffee, tea, cocoa), and many medicinal products.

The success of the angiosperms has been attributed primarily to the evolution of flowers and fruits. These structures made for more efficient means of pollination, seed dispersal, and nourishment of the embryo, as well as providing greater protection to the ovules.

25 EVOLUTION

SELF-TEST

1. Fossil evidence indicates that life has existed on earth for at least
   a 2 million years
   b 4.5 million years
   c 2 billion years
   d 4.5 billion years

2. Preservation of fossils is most likely to occur
   a where there is little rain
   b on mountain tops
   c under anaerobic conditions
   d in the tropics

3. Some cellular detail is preserved in
   a impressions
   b molds
   c casts
   d petrifactions

4. Darwin's theory of natural selection was based on
   a inheritance of acquired characteristics
   b Mendelian inheritance
   c competition between members of a species
   d chromosomal aberrations

5. One source of the heritable variations on which evolution is based is
   a mutation
   b use or disuse of an organ
   c environmental variations

6. Sexual reproduction is more important in evolution than is asexual reproduction because it results in
   a offspring that are all well adapted to a given environment
   b shorter generation times and therefore more rapid evolution
   c more offspring
   d greater variety of genotypes

7. The earliest organisms were
   a anaerobic autotrophs
   b aerobic autotrophs
   c anaerobic heterotrophs
   d aerobic heterotrophs

8. The first land plants appeared in the
   a Paleozoic Era
   b Cenozoic Era
   c Proterozoic Era
   d Mesozoic Era

9. Vascular plants are believed to have evolved from
   a green algae
   b terrestrial fungi
   c brown algae
   d bryophytes

10. Fossils of angiosperms first appeared in the
    a Cenozoic
    b Mesozoic
    c Paleozoic
    d Proterozoic

1 ______
2 ______
3 ______
4 ______
5 ______
6 ______
7 ______
8 ______
9 ______
10 ______

BASIC FACTS

Rock and soil particles deposited in layers in bodies of water eventually solidified into SEDIMENTARY ROCK. Organisms that became embedded in these deposits were often preserved as fossils. A FOSSIL is any physical indication of the existence of plants or animals in the geological past. Plant fossils may be compressions, impressions, molds, casts, or petrifactions. A COMPRESSION is a fossil flattened by the weight of layers of earth accumulated above it. An IMPRESSION is formed when a plant part (usually a leaf) falls in damp soil; the plant material gradually decays, but the impression remains as the soil hardens into rock. A MOLD is a cavity left in a rock when a plant part (usually a bulky organ such as a stem) decays after the soil in which it was embedded has hardened. If a mold becomes filled with minerals, a CAST results. A PETRIFACTION is a fossil into which dissolved minerals seeped slowly, gradually replaced the cell wall material, and became hardened into stone.

Fossils give us a record of evolution of life on the earth; the oldest rocks generally have simple organisms, the younger rocks more complex forms. The most recent rocks have fossils that most nearly resemble modern organisms. However, the fossil record is incomplete, for conditions were not always suitable for fossilization.

EVOLUTION is the gradual development of new species of plants and animals from preexisting species. In 1802 Jean LAMARCK proposed a theory of evolution by INHERITANCE OF ACQUIRED CHARACTERISTICS. He believed that changes in an organ occurring in an individual's lifetime by use or disuse are passed on to the progeny. In 1859 Charles DARWIN published his "Origin of the Species." He noted that within each species there is always some variation. Because of limitations of space and food supply, there exists a competition

(Continued on page 100)

ADDITIONAL INFORMATION

FOSSILIZATION occurs only when decomposition of dead organisms by bacterial action is retarded; this usually happens under anaerobic conditions or in water or soil that is very acid. The relative AGE of a fossil can be determined by the geological stratum in which it is embedded. A more accurate determination of age can be calculated for those fossils that still contain some of their original carbon; this method is called RADIOCARBON DATING.

In addition to mutations, evolution depends on chromosome changes and on sexual reproduction. CHROMOSOMAL ABERRATIONS such as deletions, duplications, translocations, and inversions change the genetic constitution of the organism and may also change the expression of a gene.

Asexually reproducing organisms evolve by means of mutations only and show relatively little variation from generation to generation. In SEXUALLY REPRODUCING organisms, new combinations of genes, and therefore new phenotypes, result from segregation and independent assortment. Thus there is a wide range of combinations on which natural selection can operate. As the environment changes, new species may evolve.

A difference in only one gene is not enough to produce new species; species usually differ in a number of characteristics. New species commonly arise in populations that are physically separated from other populations in the same parent species (SPACIAL ISOLATION). If environments differ in the two locations, each population usually evolves along somewhat different lines, gradually assuming those features that are best suited for its particular environment. As differences accumulate, these populations become separate species. Species that are not able to adapt to environmental changes as fast as they occur become extinct. New species may also arise by HYBRIDIZATION; some hybrid species are POLYPLOID, having more than two sets of chromosomes.

EVOLUTIONARY RELATIONSHIPS among organisms can be determined on the basis of anatomical and biochemical similarities and differences. The COMPARATIVE ANATOMY of living and extinct organisms shows that some have very similar tissues and organs and others have fewer resemblances. Similarly, COMPARATIVE BIOCHEMISTRY indicates that some organisms perform many identical or similar chemical reactions, whereas others have fewer reactions in common. Modern evolutionary trees are based on fossil, anatomical, and biochemical evidence. The fact that all organisms (except viruses) are organized into cells and the fact that all organisms have some biochemical features in common (nucleic acids as genetic material, enzymatically controlled metabolism, the ability to obtain energy by respiration) indicate that all species probably are related through common ancestors.

(Continued on page 100)

EXPLANATIONS

1. *Fossil evidence indicates that life has existed on earth for at least* 2 billion years; indirect evidence suggests that life originated even longer ago.

2. *Preservation of fossils is most likely to occur* under anaerobic conditions, usually at the bottom of a large body of water where decay is very slow.

3. *Some cellular detail is preserved in* petrifactions, but usually not in other types of fossils. Cell walls can be seen distinctly in slides of many petrifactions; rarely protoplasmic organelles are preserved.

4. *Darwin's theory of natural selection was based on* competition between members of a species and the survival of those individuals best adapted to the environment. Darwin did not know about Mendel's work or about genes and chromosomes.

5. *One source of heritable variations on which evolution is based is* mutation. Chromosomal changes are another source. Environmental variations, with a few exceptions, do not cause new heritable variations to occur, but they do determine which individuals are likely to survive and therefore which variations are preserved.

6. *Sexual reproduction is more important in evolution than is asexual reproduction because it results in* greater variety of genotypes. Some of the offspring are likely to be better adapted to the environment than their parents.

7. *The earliest organisms were* anaerobic heterotrophs. At the time they arose the earth's atmosphere had no free oxygen, and the organisms used organic compounds of nonbiological origin as their energy source.

8. *The first land plants appeared in the* Paleozoic Era about 400 million years ago. Before this happened, all life, both plant and animal, was aquatic.

9. *Vascular plants are believed to have evolved from* green algae that had migrated from the seas to the land.

10. *Fossils of angiosperms first appeared in the* Mesozoic. Angiosperms were the last of the major groups of plants to evolve.

Answers

Answer	No.
c	1
c	2
d	3
c	4
a	5
d	6
c	7
a	8
a	9
b	10

(STRUGGLE FOR EXISTENCE), and only those individuals best adapted to their environment because of favorable variations will survive and reproduce (SURVIVAL OF THE FITTEST). If these variations are inherited, each generation produces proportionately more offspring well adapted to a given environment (NATURAL SELECTION). Natural selection produces new species.

Darwin could not explain how variations arise or how they are inherited. It is now known that inherited characteristics are controlled by the genetic material, DNA. A variation arises as a MUTATION, or a change in a DNA molecule. A gene bearing a mutation that confers an advantage eventually becomes widespread throughout a population, and one that is disadvantageous becomes rare. Mutations are thought to occur at random; some are caused by radiations or certain chemicals.

Geologists divide the time since the formation of the earth into five GEOLOGICAL ERAS: the ARCHEOZOIC, extending from the formation of the earth to 2 billion years ago; the PROTEROZOIC (or PRECAMBRIAN), from 2 billion to 600 million years ago; the PALEOZOIC, from 600 million to 185 million years ago; the MESOZOIC, from 185 million to 60 million years ago; and the CENOZOIC, from 60 million years ago to the present. Simple living things are assumed to have existed during part of the Archeozoic. Algae and bacteria appeared in the Proterozoic. Early in the Paleozoic, algae were the dominant plants; about the middle of the era lycopods, horsetails, and seed ferns formed large forests. By the end of the Paleozoic these forests had disappeared, but the early gymnosperms were present. During the Mesozoic the gymnosperms became the dominant vegetation on land but were replaced by the angiosperms that suddenly arose during this era.

No organisms have ever been demonstrated to develop from anything but other organisms, yet it is believed that the FIRST ORGANISMS arose spontaneously. It is now hypothesized that about two billion years ago the earth's atmosphere contained methane (CH_4), ammonia (NH_3), and water vapor, but no free oxygen and little or no carbon dioxide. Methane, ammonia, and water contain the four most common elements in living things. With the energy of ultraviolet radiation from the sun, these inorganic compounds reacted to form both simple and complex organic compounds that were washed by rains into lakes and oceans where some of them became organized into living cells. Other organic compounds were used as a food supply by the new nonphotosynthetic organisms; foods were respired anaerobically, for no free oxygen existed. Respiration released carbon dioxide into the oceans and the atmosphere. Mutant forms arose that were able to utilize carbon dioxide in the presence of light and to reduce it to organic food molecules; these photosynthetic organisms freed living things from dependence on a dwindling food supply and produced free oxygen for the first time. Other mutations permitted organisms to utilize free oxygen in aerobic respiration. Because aerobic respiration releases more of the energy stored in foods than does anaerobic respiration, larger and more complex organisms could then develop. The presence of free oxygen also caused the formation in the upper atmosphere of a layer of ozone (O_3) which absorbed most of the harmful ultraviolet radiation from the sun. Before the development of this layer, organisms were confined to the oceans and lakes where the water, which also absorbs ultraviolet radiation, protected them. With this change in the atmosphere plants, and later animals, could colonize the land. Natural nonbiological synthesis of organic compounds today is unlikely because of the change in composition of the atmosphere.

From the first heterotrophic organisms arose the bacteria and the various divisions of algae. The fungi, which are nonphotosynthetic, probably developed from the algae. The Bryophyta and Tracheophyta are separate lines that evolved from green algae that may have become adapted to life on land. The earliest tracheophytes were probably simple Psilopsida resembling *Rhynia*; from them the modern Psilopsida, Lycopsida, Sphenopsida, and Pteropsida evolved as separate lines. The earliest of the Pteropsida were the ferns from which arose the seed ferns and the cycads and possibly the conifers, although the latter may have evolved separately from the Psilopsida. The origin of the angiosperms is unknown, but they are assumed to have arisen from early gymnosperms, possibly the seed ferns.

26 PLANT ECOLOGY

SELF-TEST

1. The largest natural biological community is called a
   a megacommunity
   b consociation
   c formation
   d biological system

2. The last community in a succession is called a(n)
   a ecosystem
   b climax community
   c ecotone
   d seral community

3. A xerophyte is a
   a plant accidentally introduced into a new region
   b parasite on an epiphyte
   c pioneer plant on land newly exposed by retreating glaciers
   d plant adapted to growing in arid regions

4. Succulents would most likely be found in
   a tropical rain forests
   b deciduous forests
   c deserts
   d tundras

5. Most of the world's coniferous forests occur in
   a Australia and the East Indies
   b subtropical areas bordering the tropical rain forests
   c subarctic areas of North America and Eurasia
   d a broad band extending from the Middle East to southeastern Asia

6. Grasslands usually occur in regions having an annual rainfall of about
   a 5 to 10 inches
   b 10 to 30 inches
   c 30 to 40 inches
   d 40 to 70 inches

7. Tundra is characterized by
   a trees and a long growing season
   b trees and a short growing season
   c small plants and a long growing season
   d small plants and a short growing season

8. Grassland with scattered trees is known as
   a prairie
   b pampa
   c steppe
   d savannah

9. The organisms at the base of a food chain are
   a photosynthetic plants
   b carnivores
   c saprophytic plants
   d herbivores

10. In general, a plant most likely to survive in areas where temperature is high, relative humidity low, and wind prevalent would have
    a large, broad leaves
    b small, narrow leaves
    c large intercellular air spaces
    d reduced palisade tissue

1 ______
2 ______
3 ______
4 ______
5 ______
6 ______
7 ______
8 ______
9 ______
10 ______

BASIC FACTS

ECOLOGY is the study of the interrelationships of organisms and their environments. An ECOSYSTEM is any natural group of plants and animals and their environment. Plants are affected by climatic, soil (edaphic), and biotic factors. CLIMATIC FACTORS include light, temperature, atmospheric gases, wind, and precipitation. SOIL FACTORS are the minerals, water, and air spaces in the soil, soil pH, and soil texture. BIOTIC FACTORS include relationships with animals and other plants.

A COMMUNITY is a group of organisms living together in an area with natural boundaries. These boundaries are determined by climatic and geological factors. The largest biological community is the FORMATION, which has a characteristic type of vegetation. A CLIMAX COMMUNITY is a stable community that ordinarily undergoes little or no change as long as the climatic and geological factors remain the same. Each formation has a characteristic climax community. The most common species of plants in a formation are the DOMINANT species; they modify the environment enough to control the populations of other plants and animals.

A PRIMARY SUCCESSION occurs when newly exposed land appears, as when a retreating glacier uncovers bare rock or a lake is filled; a temporary community occupies the area and changes the environment in such a way that another temporary community can occupy the area. The last community to develop in the area is the climax community. The particular plants in a succession are determined in part by the formation of which it is a part and in part by the condition of the newly exposed land. If a climax community is disturbed, as by fire or by lumbering, a SECONDARY SUCCESSION occurs. Where two formations meet there is an ECOTONE area that has some features of each of the formations.

(Continued on page 104)

ADDITIONAL INFORMATION

BIOTIC FACTORS affecting plants may be harmful or beneficial. Where crowding occurs there is COMPETITION between plants of the same or different species for water, minerals, or light. EPIPHYTES are nonparasitic plants that grow upon other plants and are not anchored in the soil; CLIMBERS (e.g., WOODY LIANAS) are rooted in the soil but grow over other plants. Both epiphytes and climbers damage other plants by shading them.

In areas undisturbed by man the plants and animals maintain a fairly stable balance. A limited food supply limits the numbers of predators that can survive, and the activities of predators limit the numbers of organisms on which they feed. Anything that upsets the balance usually affects more than one species. A few years of irresponsible lumbering, farming, or pest control can destroy an entire biological community and its topsoil, the formation of which required hundreds of years. CONSERVATION is a field of research which seeks to learn how we can use natural resources productively but without causing irreparable harm to the ecosystem.

Plants enter into a number of MUTUALISTIC relationships. Pollinating insects obtain food from flowers, and the plants receive the advantages of cross-pollination. Nitrogen-fixing bacteria and mycorhizal fungi depend on their host plants for food; nitrogen fixers provide nitrogenous compounds to their hosts, and mycorhizal fungi absorb minerals from the soil. Some animals disseminate seeds symbiotically, eating fruits and passing viable seeds.

The major FORMATIONS of the world are tundra, coniferous forest, temperate deciduous forest, grassland, savannah, desert, tropical rain forest, and tropical deciduous forest. ARCTIC TUNDRA occupies the northernmost areas of North America and Eurasia. The dominant vegetation consists of mosses and lichens; low herbs and shrubs are present, but no trees occur. The upper soil is usually wet and boggy during the short growing season of two to three months and is frozen for the rest of the year; permafrost underlies the upper soil. ALPINE TUNDRA occurs on the tops of tall mountains throughout the world.

CONIFEROUS FORESTS occupy a broad band around the world immediately south of the Arctic tundra; this forest is called the BOREAL FOREST. Coniferous forests also occur below alpine tundras on mountains. Most of the trees are evergreen with needle-shaped leaves; spruce, fir, larch, and pine are common. The winters in most coniferous forests are very cold, and the summers usually cool. The boreal forest occupies areas that were once glaciated and now have many lakes and bogs formed by glacial action. Rainfall is slight, but because evaporation is slow, enough water remains to support a forest.

TEMPERATE DECIDUOUS FORESTS occupy much of the United States east of the Mississippi, much of continental

(Continued on page 104)

EXPLANATIONS

1. *The largest natural biological community is called a* formation; it may cover thousands or millions of square miles.

2. *The last community in a succession is called a* climax community, which continues to reproduce itself as long as climatic and geological conditions remain unchanged.

3. *A xerophyte is a* plant adapted to growing in arid regions. Xerophytes are typical of deserts, but evergreen plants that retain their leaves in dry winter air have some characteristics of xerophytes.

4. *Succulents would most likely be found in* deserts. Succulents conserve water and can survive long dry periods.

5. *Most of the world's coniferous forests occur in* subarctic areas of North America and Eurasia. In general, coniferous forests occupy colder areas of the world than do deciduous forests.

6. *Grasslands usually occur in regions having an annual rainfall of about* 10 to 30 inches. Annual rainfall less than 10 inches is not enough to support most grasses, and forests grow where rainfall exceeds 30 inches a year.

7. *Tundra is characterized by* small plants and a short growing season, the ground being frozen most of the year.

8. *Grassland with scattered trees is known as* savannah. Savannahs occur in subtropical and tropical areas.

9. *The organisms at the base of a food chain are* photosynthetic plants; they manufacture the food used by other organisms in the chain. Saprophytes are the last organisms in the chain; they use the last of the energy-rich organic compounds in the chain and convert them to low-energy inorganic compounds that only plants can use.

10. *In general, a plant most likely to survive in areas where temperature is high, relative humidity low, and wind prevalent would have* small, narrow leaves that tend to lose only a small amount of water by transpiration.

Answers

c	1
b	2
d	3
c	4
c	5
b	6
d	7
d	8
a	9
b	10

XEROPHYTES are plants that are able to live in arid regions; they have adaptations that enable them to conserve water and resist desiccation. HYDROPHYTES live partially or completely submerged in water, and MESOPHYTES occupy areas with intermediate amounts of moisture.

The four major TYPES OF VEGETATION are forest, grassland, desert, and tundra. Temperature and rainfall are the chief factors that determine which type of vegetation grows in a given area. FORESTS usually occur in areas having more than 30 inches of rainfall each year, GRASSLANDS in areas with 10 to 30 inches a year, and DESERTS in areas with less than 10 inches a year. TUNDRAS are located where the subsoil is permanently frozen. A fifth type of vegetation, the SAVANNAH, is grassland with a few scattered trees.

A FOOD CHAIN is a sequence of organisms in which each organism is placed below the one that eats it. A food chain represents the direction of energy flow in the ecosystem. PHOTOSYNTHETIC PLANTS occupy the base of every food chain, for they are the only organisms that can manufacture food. At the second level are HERBIVOROUS ANIMALS that eat plants. The third level is occupied by CARNIVOROUS ANIMALS THAT EAT HERBIVORES, and the fourth by CARNIVORES THAT EAT OTHER CARNIVORES. PARASITES may occur at any level above the first one. At the top level are SAPROPHYTES that cause the decay of dead organisms and convert their bodies into inorganic compounds used by plants. A food chain is actually more of a web than a straight chain, for one animal can occupy more than one position in it. Some animals are OMNIVOROUS, eating both plants and animals. Food chains occur in the sea as well as on land.

Europe north of the Alps, the eastern part of Asia, and scattered areas in the southern hemisphere. Winters are cold but not severe. Common climax trees in different parts of the deciduous forest in eastern United States are oak, hickory, maple, beech, and basswood.

GRASSLAND occurs between the Rocky Mountains and the Mississippi River in the United States, where it is called PRAIRIE, and in a broad band across central Asia, where it is called STEPPE. Smaller areas of grasslands, including the PAMPAS of South America, occur in other parts of the world. The dominant grasses range from a few inches tall in dry grasslands to ten feet tall in wetter areas. Precipitation is generally low, but soils are usually dark and fertile and make good farming or range land. SAVANNAHS are common in tropical and subtropical parts of Africa and South America and in parts of Australia and India.

Most of the world's DESERTS lie near 30° N latitude or 30° S latitude. The largest desert extends in a broad band across all of northern Africa, all of Arabia, and most of Asia at this latitude. Much of southwestern United States is desert. Rainfall in deserts is low and the evaporation rate is high. Some deserts have almost no plants; others have a number of species, though the vegetation is not so dense as in other formations. Desert plants are mainly small-leaved, deciduous shrubs or succulents such as cacti.

TROPICAL RAIN FORESTS occur in tropical areas with high rainfall: the Amazon basin of South America, Central America, the Congo basin of Africa, southeast Asia, and the East Indies. The temperature and the relative humidity are always high. These forests have great numbers of species of trees; most of them are evergreen angiosperms. TROPICAL DECIDUOUS FORESTS occupy those areas of the tropics that have dry seasons during which the trees are devoid of leaves.

The amount of energy available to organisms depends on their positions in the FOOD CHAIN. Because most of the food manufactured by plants or eaten by an animal is oxidized in respiration or is converted to inedible parts, only a small amount of this food and a correspondingly small amount of energy is available to the next organism in the chain. Thus of the light energy that plants convert into chemical energy only 1/10 is available to herbivores, only 1/100 to carnivores feeding on herbivores, and only 1/1000 to the top carnivores. A given acreage of land supporting a certain quantity of plant material can support a smaller quantity of herbivores and a still smaller quantity of carnivores. Human populations that live primarily on meat require much more land to support them than do populations that eat only plants. In overpopulated countries most people are of necessity vegetarians.

Final Examination
Dictionary-Index

FINAL EXAMINATION

Questions

DIRECTIONS: Write your answer (a, b, c, d, etc., or the missing term) on the numbered lines in the margins. To check your answers, turn to page 108. Study the explanations for any questions you missed.

1 ____________
2 ____________
3 ____________
4 ____________
5 ____________
6 ____________
7 ____________
8 ____________
9 ____________
10 ____________
11 ____________
12 ____________
13 ____________
14 ____________
15 ____________
16 ____________

1. All living plant cells perform
   a photosynthesis
   b respiration
   c transpiration
   d meiosis

2. Most biochemical reactions differ from those in the nonliving world in
   a requiring energy
   b releasing energy
   c being enzymatic
   d being spontaneous

3. Two compounds characteristic of plants but not of animals are
   a pyruvic acid and glucose
   b glucose and cellulose
   c cellulose and starch
   d starch and pyruvic acid

4. Thick secondary cell walls may be found in
   a parenchyma cells
   b storage cells
   c dividing cells
   d dead cells

5. Growth in width of stems and roots is due to the activity of
   a primary meristems
   b secondary meristems
   c intercalary meristems
   d apical meristems

6. Cambium is found primarily in ____________ (monocots or dicots?).

7. A compound that transfers energy from energy-releasing reactions to energy-requiring reactions is
   a tRNA
   b ATP
   c IAA
   d AMP

8. The conversion of carbon dioxide to organic compounds ____________ (requires or releases?) energy.

9. The turgor pressure of a cell could be raised by placing it in a solution of ____________ (lower or higher?) diffusion pressure.

10. Nucleic acids are found in
    a the nucleus only
    b the chromosomes only
    c the cytoplasm only
    d both nucleus and cytoplasm

11. Ordinarily two genetically identical daughter cells are produced by ____________ (mitosis or meiosis?).

12. The abundant genotypic and phenotypic variety of sexually reproducing organisms is due largely to ____________ (mitosis or meiosis?).

13. The appearance of a plant due to its genetic constitution is called its ____________ (genotype or phenotype?).

14. The type of RNA that carries amino acids to the ribosomes is called ____________ (messenger or transfer?) RNA.

15. When a stem is unilaterally illuminated, the shaded side has a ____________ (higher or lower?) concentration of auxin than the illuminated side.

16. Which of the following is not a thallophyte?
    a lichen
    b alga
    c slime mold
    d moss

In the following five questions, write the letter(s) of the compound(s) in Column B that is (are) part of the complex compound in Column A.

Column A		*Column B*
17. ATP	a	monosaccharide
18. starch	b	phosphate
19. fats	c	purine
20. DNA	d	glycerol
21. proteins	e	nitrate
	f	uracil
	g	pyrimidine
	h	fatty acid
	i	amino acid

In the following seven questions, write the letter of the function in Column B that is associated with the item in Column A.

Column A		*Column B*
22. chloroplasts	a	food translocation
23. chromatids	b	photoperiodism
24. ribosomes	c	phototropism
25. stomata	d	respiration
26. mitochondria	e	digestion
27. phloem	f	photosynthesis
28. phytochrome	g	abscission
	h	nastic movements
	i	cell division
	j	exchange of gases
	k	protein synthesis

29. The sporophyte generation ordinarily is ___________ (haploid or diploid?).

30. Algae and bryophytes are ___________ (autotrophic or heterotrophic?) plants.

31. The gametophyte is the dominant generation in
   a ferns
   b conifers
   c mosses
   d brown algae

32. Which of the following do not require external water for fertilization?
   a mosses
   b cycads
   c ferns
   d liverworts

33. Archegonia are absent in
   a bryophytes
   b club mosses
   c conifers
   d monocots

34. A plant in which eggs and sperms arise from the same gametophyte is
   a isogamous and homothallic
   b isogamous and heterothallic
   c heterogamous and heterothallic
   d heterogamous and homothallic

35. Double fertilization occurs in
   a angiosperms
   b gymnosperms
   c algae
   d fungi

36. The most primitive living vascular plants are
   a psilopsids
   b lycopods
   c sphenopsids
   d cycads

37. The order of organisms in a typical food chain is
   a diatoms → crustacea → fish → seal → polar bear → bacteria
   b bacteria → seal → diatoms → polar bear → fish → crustacea
   c crustacea → seal → fish → polar bear → bacteria → diatoms
   d polar bear → seal → diatoms → bacteria → fish → crustacea

17 ________
18 ________
19 ________
20 ________
21 ________
22 ________
23 ________
24 ________
25 ________
26 ________
27 ________
28 ________
29 ________
30 ________
31 ________
32 ________
33 ________
34 ________
35 ________
36 ________
37 ________

Answers

b	1
c	2
c	3
d	4
b	5
dicots	6
b	7
requires	8
higher	9
d	10
mitosis	11
meiosis	12
phenotype	13
transfer	14
higher	15
d	16

EXPLANATIONS

1. *All living plant cells perform* respiration, which releases the energy stored in foods.

2. *Most biochemical reactions differ from those in the nonliving world in* being enzymatic. A few reactions in cells are spontaneous, but even most of the energy-releasing reactions do not occur in the absence of enzymes.

3. *Two compounds characteristic of plants but not of animals are* cellulose and starch. Most plant cells are surrounded by cellulose cell walls; most animal cells lack walls. Plants commonly store starch as a food reserve, but animals store glycogen. Glucose and pyruvic acid can be found in nearly all living cells; pyruvic acid is an intermediate compound in the respiration of glucose.

4. *Thick secondary cell walls may be found in* dead cells. At maturity tracheids, vessels, xylem and phloem fibers, and sclerenchyma have secondary walls but no protoplasts.

5. *Growth in width of stems and roots is due to the activity of* secondary meristems, particularly the vascular cambium.

6. *Cambium is found primarily in* dicots, especially in perennials. Only a few monocots have cambium.

7. *A compound that transfers energy from energy-releasing reactions to energy-requiring reactions is* ATP, which is probably the most common energy-transferring compound in living cells.

8. *The conversion of carbon dioxide to organic compounds* requires *energy.* The most commonly used energy is the light used by photosynthetic plants.

9. *The turgor pressure of a cell could be raised by placing it in a solution of* higher *diffusion pressure*; water enters the cell, and the turgor pressure inside the cell increases.

10. *Nucleic acids are found in* both nucleus and cytoplasm. The genetic material, DNA, is confined largely to the chromosomes in the nucleus. Ribosomal RNA and transfer RNA occur in the cytoplasm, and messenger RNA carries the genetic code from the nucleus to the cytoplasm.

11. *Ordinarily two genetically identical daughter cells are produced by* mitosis. Somatic cells of a plant are genetically identical.

12. *The abundant genotypic and phenotypic variety of sexually reproducing organisms is due largely to* meiosis, during which independent assortment of chromosomes occurs. Asexually produced organisms usually are genetically identical to their parents.

13. *The appearance of a plant due to its genetic constitution is called its* phenotype. The phenotype can be influenced by the environment as well as by the genotype.

14. *The type of RNA that carries amino acids to the ribosomes is called* transfer *RNA*. Messenger RNA carries the genetic code.

15. *When a stem is unilaterally illuminated, the shaded side has a* higher *concentration of auxin than the illuminated side.* The difference in concentration produces unequal growth rates and causes the stem to bend toward the light.

16. A moss is not a thallophyte but is a nonvascular member of the subkingdom Embryophyta.

17. *ATP* consists of a monosaccharide (ribose), three phosphates, and a purine (adenine).

18. *Starch* consists of repeating monosaccharide (glucose) units.

19. *Fats* consist of glycerol combined with three fatty acids; the fatty acids may be the same or different.

20. *DNA* consists of repeating units called nucleotides, each of which contains a monosaccharide (deoxyribose), a phosphate, and either a purine (adenine or guanine) or a pyrimidine (cytosine or thymine).

21. *Proteins* consist of repeating amino acid units.

22. *Chloroplasts* are the site of photosynthesis in nearly all photosynthetic plants.

23. *Chromatid* formation is associated with cell divisions including mitosis and meiosis.

24. *Ribosomes* are the site of protein synthesis.

25. *Stomata* permit exchange of gases (especially carbon dioxide, oxygen, and water vapor) with the atmosphere.

26. *Mitochondria* are the site of the Krebs cycle portion of aerobic respiration.

27. *Phloem* transports food both upward and downward in vascular plants.

28. *Phytochrome* is the photoreceptive pigment in photoperiodic phenomena in plants.

29. *The sporophyte generation ordinarily is* diploid; the gametophyte generation ordinarily is haploid.

30. *Algae and bryophytes are* autotrophic *plants*, as are tracheophytes. The fungi lack chlorophyll and are heterotrophic.

31. *The gametophyte is the dominant generation in* mosses. The sporophyte is the dominant generation in the conifers and some brown algae; in other brown algae the sporophyte and gametophyte generations are identical vegetatively.

32. Cycads do not require external water for fertilization. Pollen grains are transported by wind to the ovulate cones where they release their sperms.

33. *Archegonia are absent in* monocots and dicots. Gametophytes of angiosperms are so reduced that they have neither archegonia nor antheridia.

34. *A plant in which eggs and sperms arise from the same gametophyte is* heterogamous and homothallic. Isogamous species produce gametes that are alike in appearance. Heterothallic species produce male and female gametes or gametes of opposite mating types on different gametophytes.

35. *Double fertilization occurs in* angiosperms, the only plants to have a polyploid endosperm in their seeds.

36. *The most primitive living vascular plants are* psilopsids. *Psilotum* has a rhizome with dichotomously branched aerial stems that are photosynthetic. There are no roots, but the rhizome has rhizoids.

37. *The order of organisms in a typical food chain is* diatoms → crustacea → fish → seal → polar bear → bacteria. The photosynthetic plants are the base of the chain. The amount of energy available to organisms decreases with every step upward in the chain.

Answers

Answer	No.
a,b,c	17
a	18
d, h	19
a, b, c, g	20
i	21
f	22
i	23
k	24
j	25
d	26
a	27
b	28
diploid	29
autotrophic	30
c	31
b	32
d	33
d	34
a	35
a	36
a	37

DICTIONARY-INDEX

Abscission: the dropping of leaves, fruits, or other organs along a preformed weak zone called the abscission zone, 16, 58

Abscission zone: a zone of weak cells found at the base of leaves, fruits, and other organs that drop as the cells disintegrate, 16

Accessory fruit: a fruit that includes the floral tube or stem tissues, 48

Achene: a one-seeded indehiscent fruit in which the seed is attached to the pericarp only by its funiculus, 48

Actinomorphic: referring to radially symmetrical (regular) flowers, 94

Actinomycetes: a group of funguslike bacteria, 70

Active absorption: absorption of water due to osmosis, 32

Adenosine diphosphate: a precursor of adenosine triphosphate, 18

Adenosine triphosphate: a common energy-transferring compound in cells, 18

ADP: adenosine diphosphate, 18, 20, 24

Adventitious root: a root that arises from an organ other than a root, 26, 46, 82, 86, 88

Aecium: a stage in the life cycle of wheat rust, 76

Aerobe: an organism that requires free oxygen, 70

Aerobic respiration: oxidation of food in the presence of free oxygen, 22, 100

Afterripening: a series of physiological changes that occur within seeds before they germinate, 48

Aggregate-accessory fruit: an aggregate fruit that includes the receptacle on which the flower grew, 48

Aggregate fruit: a fruit that develops from several simple pistils of a single flower, 48

Aleurone: protein-rich outer layer of the endosperm of cereal grains, 48

Alga: a member of the Thallophyta that possesses chlorophyll and manufactures food by photosynthesis, 64, 66, 76, 100

Algology: study of algae, 2

Allele: one of two or more alternate forms of a gene, 54

Alpine tundra: 102

Alternate leaf arrangement: arrangement of leaves in which only one leaf occurs at a node, 38

Alternation of generations: alternation of gametophyte and sporophyte generations typical of the Embryophyta and of some Thallophyta, 62, 68, 78, 88

Amino acid: a simple organic compound having an acid (—COOH) group and an amine (—NH_2) group, 8, 52

Anabolism: metabolic reactions that synthesize complex compounds from simpler ones; e.g., photosynthesis, synthesis of starch from glucose, 2

Anaerobe: an organism that does not require free oxygen, 70

Anaphase: a phase of mitosis in which identical chromatids separate and move to opposite poles of the cell, 10, 12, 50

Androecium: the stamens of a flower collectively, 42

Angiospermae: the class of flowering plants, 94, 100

Annulus: a row of cells that assists in the forcible discharge of spores from a fern sporangium, 86

Anther: the pollen-producing part of a stamen, 42, 44

Antheridiophore: a stalk bearing antheridia, 78

Antheridium: sperm-producing organ of nonseed plants, 66, 74, 76, 78, 82, 88

Anthoceros: 80

Anthocerotae: a class of the Bryophyta consisting of the hornworts, 80

Anthocyanins: water-soluble pigments ranging in color from red to blue, 16

Antibiotic: a substance produced by one organism that inhibits the growth of other organisms, 70

Antipodal cells: the cells of an embryo sac opposite the micropyle, 44

Apical dominance: inhibition of the growth of lateral buds by the apical bud, 60

Apical meristem: meristem at the apex of a stem or root; apical meristems produce primary tissues and provide increase in length, 12, 26, 38, 82, 86, 90

Apogamy: formation of a haploid sporophyte from vegetative cells of a gametophyte, 88

Apospory: formation of a diploid gametophyte from vegetative cells of a sporophyte, 88

Apothecium: a cup-shaped ascocarp of some Ascomycetes, 76

Archegoniophore: a stalk bearing archegonia, 78

Archegonium: multicellular female reproductive organ of members of the subkingdom Embryophyta exclusive of the angiosperms, 78, 82, 88, 90, 92

Archeozoic era: 100

Arctic tundra: 102

Ascocarp: fruiting body of some Ascomycetes, 76

Ascogonium: female reproductive organ of some Ascomycetes, 76

Ascomycetes: a class consisting of the sac fungi, 74, 76

Ascospore: sexual spore of Ascomycetes, 74, 76

Ascus: saclike sexual sporangium of the Ascomycetes, 74, 76

Asexual reproduction: reproduction not involving union of gametes, 16, 28, 40, 62, 66, 78, 80, 88

Asexual spore: a spore produced by mitosis, 62, 76

Atom: the smallest unit of an element that can take part in a chemical reaction, 6

ATP: adenosine triphosphate, 18, 20, 22, 24, 36

Autonomic movement: spontaneous movement controlled by internal stimuli, 60

Autotrophic: capable of synthesizing food from inorganic compounds, 66, 72

Autumn coloration: 16

Auxin: a plant growth hormone, either natural or artificial, 58, 60

Axil: the angle formed by the upper surface of a leaf and the stem, 38

Axile placentation: placentation in which the placentas of a compound ovary meet in the center of the ovary, 94

Axillary bud: lateral bud, formed in an axil, 4, 38

Bacillariophyceae: diatoms, 68

Bacillus: a rod-shaped bacterial cell, 70

Back cross: test cross, a genetic test to determine homozygosity or heterozygosity of an individual, 56

Bacteria: microscopic, single-celled plants without true nuclei, 70, 100
Bacteriology: the study of bacteria, 2
Bacteriophage: a virus that parasitizes bacteria, 72
Bark: all tissues external to the cambium of woody stems, 40
Basal placentation: placentation in which a single ovule arises from the base of a simple ovary, 96
Basidiocarp: fruiting body of some Basidiomycetes, 76
Basidiomycetes: a class consisting of the club fungi, 74, 76
Basidiospore: sexual spore of Basidiomycetes, 76
Basidium: a structure that bears basidiospores, 74, 76
Bean: 46
Bennettitales: an order of extinct gymnosperms, 92
Berry: a many-seeded fleshy fruit derived from a compound ovary, 48
Biennial: a plant that completes its life cycle in two growing seasons, 60
Binomial: a name consisting of a generic and a species name assigned to all organisms, 62, 64
Biochemistry: the study of chemical reactions occurring within organisms, 2, 98
Blade: flattened portion of a leaf, 14
Bleeding: exudation of water from wounds due to high root pressure, 30
Blending: incomplete dominance; the contribution of two allelic genes to a phenotypic expression, 54
Blue-green algae: a group of algae comprising the division Cyanophyta, 66, 68, 72
Body cell: the cell of a gymnosperm pollen grain that divides into two sperm cells, 90
Boreal forest: 102
Botany: the science concerned with the study of plants, 2
Bract: a modified leaf below a flower or inflorescence, 44
Branch: 38
Branch root: a root arising from another root; a secondary root, 26
Brown algae: marine algae comprising the division Phaeophyta, 68
Bryology: the study of bryophytes, 2
Bryophyta: a division including the nonvascular members of the embryophyta; liverworts, hornworts, and mosses, 64, 78, 100
Bud: in vascular plants, a very young branch consisting of a stem tip, unexpanded leaves, and bud scales, 4, 38, 60; in yeasts, a small outgrowth of a cell that develops into a new cell, 76
Bud scale: a modified leaf protecting a bud, 16, 38
Bulb: a short, vertical underground stem with large, fleshy leaves, 40
Bundle sheath: a parenchymatous tissue surrounding vascular bundles of some leaves, 14

Calamites: 84
Calyptra: cap of a moss capsule derived from the archegonium, 80
Calyx: all the sepals of a flower collectively, 42, 94
Calyx tube: floral tube or hypanthium; the fused bases of sepals, petals, and stamens, 94
Cambium: a lateral (secondary) meristem, either vascular or cork cambium, 26, 28, 40, 90
Capillary water: water that soil can hold against gravity and that is available to plants, 34
Capsule: a dehiscent fruit derived from a compound ovary, 48; spore-producing structure of the bryophytes, 78; slime layer surrounding some bacterial cells, 70
Carbohydrate: an organic compound composed of carbon, hydrogen, and oxygen with hydrogen and oxygen atoms in a ratio of 2:1, 6, 8
Carbon cycle: the cycle of carbon compounds within and among organisms and the environment, 70
Carotene: an orange pigment associated with chlorophyll in chloroplasts, 16, 66, 68, 78
Carpel: the megasporophyll of a flowering plant, 42, 94
Carpogonium: female reproductive organ of red algae, 68
Carpospore: a spore produced on the carpogonial filaments of some red algae, 68
Caryopsis: a one-seeded indehiscent fruit in which the seed coat is fused at all points to the pericarp; a grain, 48
Casparian strip: a suberized strip found in the upper, lower, and radial walls of endodermal cells, 28
Cast: a fossil formed by the filling of a mold with minerals, 98
Catabolism: metabolic reactions that convert complex compounds into simpler ones; e.g., digestion, respiration, 2
Catkin: a spike that contains only imperfect flowers, 42
Cell: the fundamental living unit usually organized into cytoplasm and nucleus, 10
Cell plate: a thin membrane that forms across a cell in telophase, dividing the cell into daughter cells, 12
Cell wall: a wall composed mainly of cellulose and surrounding the protoplast of most plant cells, 2, 10, 70, 74
Cellulose: one of the major constituents of plant cell walls, a long-chain polysaccharide composed of repeating glucose subunits, 6
Cenozoic era: 100
Central body: the region of a bacterial cell that contains genetic material, 66, 70
Centromere: kinetochore; the region of a chromosome to which spindle fibers attach, 10
Chemical bond: a bond between two atoms of a molecule, 8
Chemical equation: a short method of writing a chemical reaction, 6
Chemical formula: an abbreviation for the composition of a compound, representing one molecule, 6
Chemical symbol: an abbreviation for the name of an element, representing one atom, 6
Chemosynthesis: synthesis of food using energy obtained from the oxidation of inorganic compounds, 72
Chemotropism: a growth movement in response to chemical stimuli, 60
Chitin: a horny substance present in the cell walls of many fungi, 74
Chlorophyll: a green pigment that absorbs the light energy used in photosynthesis, 18, 66, 68, 78
Chlorophyta: a division comprising the green algae, 68
Chloroplast: a specialized body in the cytoplasm in which photosynthesis takes place, 10, 16, 66, 68, 78
Chlorosis: loss of chlorophyll, 34
Chromatid: one of the two identical portions of a replicated chromosome, 10, 50, 52
Chromoplast: a plastid containing pigments other than green, 10

Chromosomal aberrations: a change in the structure of a chromosome, 52, 98

Chromosome: a long, threadlike body confined to the nucleus of the cell and containing genetic material, 10, 12, 50, 52, 54

Chromosome map: a map indicating the location of genes in a chromosome, 54

Chrysophyceae: a class consisting of golden-brown algae, 68

Chrysophyta: a division of algae including diatoms, yellow-green algae, and golden-brown algae, 68

Citric acid cycle: Krebs cycle, 22

Cladophyll: a leaflike stem, 40

Class: a group of closely related orders, 62

Classification of plants: 64

Clay: soil with particles of less than 0.002 mm. diameter, 34

Cleistothecium: a closed, spherical ascocarp of some Ascomycetes, 76

Climax community: the final stable community in a biological succession, 102

Climber: a plant that is rooted in soil and that climbs over other plants, 102

Club fungi: fungi comprising the class Basidiomycetes and characterized by production of sexual spores on basidia, 74

Club mosses: nonseed vascular plants comprising the subdivision Lycopsida, 82

CoA: coenzyme A, 22

Coal formation: 84

Coccus: a spherical bacterial cell, 70

Codon: a set of three nucleotides in a DNA molecule that codes for one amino acid in a protein molecule, 52

Coenocyte: a multinucleate organism with no cross walls dividing it into separate cells, 66, 68, 74

Coenzyme A: a coenzyme involved in aerobic respiration and lipid metabolism, 22

Coleoptile: a sheath surrounding the epicotyl of grass embryos, 46

Coleorhiza: a sheath surrounding the radicle of grass embryos, 46

Collenchyma: a supporting tissue of living, thick-walled cells, 12

Colloid: a substance composed of microscopic particles large enough to remain in suspension, 8

Columella: sterile, central portion of a capsule or sporangium, 78, 80

Community: a group of plants and animals living in an area with natural boundaries, 102

Companion cell: a nucleated phloem cell closely associated with a sieve-tube element, 12, 38, 94

Complete flower: a flower that has sepals, petals, stamens, and pistils, 44, 94

Compound: a substance composed of two or more elements chemically combined in definite proportions by weight, 6

Compound leaf: a leaf in which the blade is divided into leaflets, 14

Compression: a fossil flattened by the weight of earth accumulated above it, 98

Cone: strobilus; a stem bearing sporophylls, 4, 84, 90, 94

Conidiophore: a hypha bearing conidia, 76

Conidiospore: conidium, 76

Conidium: an asexual fungal spore, usually produced in chains, 76

Coniferales: an order of woody gymnosperms having needle- or awl-shaped leaves, 90, 92, 100

Coniferous forest: a forest in which conifers are the dominant vegetation, 102

Conjugation: fusion of two isogametes, 66; in bacteria, 70

Conservation: 102

Cordaitales: an order of extinct woody gymnosperms, 92

Cork: the external tissue of stems and roots with secondary growth; secondary tissue produced by cork cambium, 12, 26, 28, 40

Cork cambium: a secondary (lateral) meristem that produces cork and phelloderm, 26, 28, 40

Corm: a short, thick, vertical, underground stem with small, scalelike leaves, 40

Corn: 46

Corolla: all the petals of a flower collectively, 42

Cortex: a primary tissue found between the epidermis and vascular tissue of stems and roots, 26, 40

Corymb: an inflorescence with a long peduncle from which branch the pedicels of the flowers; pedicels of the lower flowers are longer than those of the upper flowers, 42

Cotyledon: a seed leaf, 16, 46, 90, 96

Crista: an invagination of the inner unit membrane of a mitochondrion, 24

Crossing over: exchange of homologous portions of homologous chromosomes during meiosis, 54

Cross-pollination: pollination of the pistil of one plant by pollen from another plant, 44, 96, 102

Crystal: a nonliving cytoplasmic inclusion usually consisting of waste products, 12

Cuticle: a waxy secretion of epidermal cells, 14, 16, 82

Cyanophyta: a division comprising the blue-green algae, 68

Cycadales: an order of gymnosperms with large, pinnately compound leaves, 90, 92, 100

Cyclic photophosphorylation: one of the light reactions in photosynthesis; synthesis of ATP using light energy, 18

Cyme: an inflorescence in which the terminal flower opens first, 42

Cytochrome: an electron-transferring compound in cells, 18, 22

Cytokinesis: division of cytoplasm usually accompanying mitosis, 10, 12, 74

Cytology: the study of cells, 2

Cytoplasm: protoplasm of a cell exclusive of its nucleus, 10, 30

Cytoplasmic inheritance: inheritance of genetic material located in the cytoplasm, 56

Dark reactions: a series of reactions in photosynthesis in which carbon dioxide is reduced to an organic compound with the ATP and hydrogen produced in the light reactions, 18, 20

Darwin, Charles: a nineteenth-century naturalist who proposed a theory of evolution based on natural selection, 98

Day-neutral plant: a plant the flowering of which is not influenced by day length, 58

Decay: decomposition of organic materials, 72

Deciduous: referring to plants that lose all their leaves once a year, 16, 92, 94, 102, 104

Dehiscent: referring to fruits that split open at maturity, 48

Deoxyribonucleic acid: genetic material of the cell, 50, 52, 72, 100

Desert: a vegetation type characteristic of arid regions of the temperate zones, 104

Deuteromycetes: an artificial class of imperfect fungi, 74

Diatoms: a group of single-celled algae comprising the class Bacillariophyceae of the division Chrysophyta, 68

Dicaryotic: in fungi, having two nuclei (from two different hyphae) per cell, 74, 76

Dichogamy: maturation of stamens and pistils of the same flower at different times, 96

Dicotyledonae: a subclass of angiosperms characterized by having two cotyledons in the seed, 94, 96

Dictyosome: Golgi apparatus, 10

Dictyostele: a siphonostele with several leaf gaps at any one level, 86

Differentially permeable membrane: a membrane that permits the passage of some substances but not others, 30

Differentiation: maturation of specialized tissues from meristematic tissues, 58

Diffuse porous: referring to wood with vessels uniformly distributed throughout the spring and summer wood, 38

Diffusion: net movement of molecules of a substance from an area of high diffusion pressure of that substance to an area of low diffusion pressure of that substance, 30

Diffusion pressure: tendency of molecules to move, 30

Digestion: conversion of complex foods into their subunits; 22

Dihybrid: heterozygous for two pairs of genes, 56

Dimorphic: having two forms; e.g., having vegetative and reproductive leaves, 86

Dioecious: having male and female flowers (or cones) on separate plants, 44, 90, 92, 96

Diphosphopyridine nucleotide: DPN; a hydrogen carrier in cells, also called nicotinamide adenine dinucleotide (NAD), 22

Diplococcus: a bacterium having spherical cells occurring in pairs, 70

Diploid: having two sets of chromosomes per nucleus, 42, 50, 62

Disaccharide: a sugar composed of two monosaccharides, 6

Dispersal of seeds and fruits: 48

Division: a group of related classes, 62

DNA: deoxyribonucleic acid, 50, 52, 72, 100

Dominant gene: a gene that produces its phenotypic effect even in the presence of its recessive allele, 54

Dominant species: a species that because of its numbers (or size) controls the population of a community, 102

Dormancy: a temporary state of seeds or buds in which growth and physiological activity are low, 48, 90

Double fertilization: in flowering plants, the fertilization of the egg and of the polar nuclei by two sperm nuclei, 44, 96

DPN: diphosphopyridine nucleotide, 22

Drupe: a one-seeded fleshy fruit derived from a simple ovary, the endocarp of which becomes hard and stony, 48

Ecology: the study of the relationships existing between plants, animals, and their environments, 2, 102

Ecosystem: any natural community of organisms interacting with each other and their physical environment, 102

Ecotone: an area having features intermediate between those of two neighboring formations, 102

Egg: nonmotile female gamete, 42, 44, 66

Elater: a cell or part of a cell that assists in the dissemination of spores in liverworts and horsetails, 78, 84

Electrons: negatively charged atomic particles, 6, 18

Element: a substance that cannot be broken down into simpler chemical substances by ordinary chemical means, 6

Embryo: a young plant retained within the archegonium or seed, 4, 44, 46, 78, 96

Embryo sac: megagametophyte of a flowering plant, 42, 44, 94, 96

Embryo sac mother cell: megaspore mother cell of a flowering plant, 42, 96

Embryophyta: the subkingdom composed of plants that have multicellular reproductive organs and retain embryos within the female reproductive organ, 64, 78

Endocarp: the inner layer of the pericarp, 46

Endodermis: a single layer of cells that surrounds the vascular tissue and pericycle of roots and some underground stems, 28

Endoplasmic reticulum: a network of double membranes in the cytoplasm, 10

Endosperm: a polyploid, nutritive tissue found in the seeds of some flowering plants, 44, 46, 96

Endosperm mother cell: a two-nucleate cell of the embryo sac, 96

Endosperm nucleus: polyploid nucleus of the endosperm cell that is formed by the fertilization of two polar nuclei by a sperm nucleus, 44

Energy: in physics, the ability to do work, 6, 18, 22, 36

Energy flow: the passage of chemical energy stored in foods from organism to organism along a food chain, 104

Enzyme: a proteinaceous catalyst that speeds metabolic reactions, 12, 22, 24

Enzyme synthesis: synthesis of catalytic proteins on ribosomes, 52

Epicotyl: the portion of a seed plant embryo above the cotyledons, including the stem tip and young vegetative leaves, 46, 90

Epidermis: outermost layer of cells of plant organs with only primary growth, 12, 16, 26

Epigynous: referring to flowers in which the sepals, petals, and stamens appear to arise from the top of the ovary, 94

Epiphyte: a plant that grows on another plant but does not parasitize it, although it may cause harm, 102

Equisetum: 82

Euglenophyta: a division of algae comprising the euglenoids, 66, 68

Eumycophyta: a division consisting of true fungi, 74

Evergreen: referring to plants that retain leaves throughout the year, 16, 94, 102, 104

Evolution: the development of new species (or other taxa) from other species by a process of gradual change, 98

Exocarp: the outer layer of the pericarp, 46

Experiment: a method of testing a hypothesis, 4

F_1: first generation of offspring, 54

F_2: second generation of offspring, 56

Family: a group of closely related genera, 62

Fascicular cambium: vascular cambium located within a vascular bundle, 40

Fat: an organic compound consisting of carbon and hydrogen in a ratio of about 1:2 and very little oxygen; a compound formed by the condensation of three fatty acid molecules with a glycerol molecule, 8, 22

Fatty acid: a simple organic compound consisting mostly of carbon and hydrogen and terminating in an acid (—COOH) group, 8

Fermentation: oxidation of a simple sugar in the absence of free oxygen, 24, 72

Ferns: a group of nonseed vascular plants comprising the class Filicineae, 86, 100

Fertilization: fusion of two gametes, especially two heterogametes, 62

Fiber: a thick-walled, elongated, strengthening cell, 12, 38

Fibrous root system: a root system in which all of the roots are about the same size, 26

Fiddlehead: a young, coiled fern leaf, 86

Field capacity: the water that a soil can hold against gravity, 34

Filament: stalk of a stamen, 42; threadlike plant body of some thallophytes, 66

Filicineae: a class of the Pteropsida that comprises the ferns, 86

Fission: cell division of a unicellular organism into two daughter cells of equal size, 70

Flagellum: a whiplike organelle by which certain cells move, 66, 70, 74, 82, 92

Floral tube: calyx tube or hypanthium; the fused bases of sepals, petals, and stamens, 42, 94

Floridean starch: food reserve of red algae, 68

Florigen: a hypothetical hormone assumed to initiate flowering in the angiosperms, 58, 60

Flower: sexual reproductive structure of the angiosperms; a short stem usually bearing sepals, petals, stamens, and one or more pistils, 42, 58, 94, 96

Follicle: a dehiscent fruit that splits open along only one side, 48

Food: organic compounds that serve as energy sources and as building material for cellular growth, 2, 8, 18, 22, 48

Food chain: a sequence of organisms arranged according to their food sources, 104

Foot: the portion of an embryo that remains within the venter and absorbs food from the gametophyte, 78, 88

Forest: vegetation type in which the dominant species are trees, 104

Formation: the largest natural biological community, 102

Fossil: any physical evidence of the existence of organisms in the geological past, 84, 92, 98

Free central placentation: placentation in which ovules are arranged on a central stalk arising from the base of the ovary, 94

Frond: leaf of a fern, 86

Fructose: a simple sugar, 6

Fruit: the ripened ovary of a flowering plant, 4, 44, 46, 94

Fucoxanthin: a brown pigment of the brown algae, 68

Fungi Imperfecti: an artificial class of imperfect fungi; Deuteromycetes, 74, 76

Fungus: a member of the Thallophyta that lacks chlorophyll and is unable to manufacture food, 64, 74, 100

Funiculus: stalk of an ovule or seed, 42, 46

Gametangium: a structure that produces gametes, 74

Gamete: a sex cell, 50, 62

Gametophyte: a plant, usually haploid, that produces gametes, 62, 68, 78, 80, 82, 84, 88, 90, 94

Gemma: a budlike structure produced by some bryophytes that may form a new plant, 78

Gene: a unit of inheritance located on a chromosome, 50, 54, 98, 100

Generative cell: in flowering plants, the cell in a pollen grain that divides into two sperm cells, 44; in gymnosperms, the cell in a pollen grain that divides into a stalk cell and a body cell, the latter producing sperms, 90

Genetic code: information coded in DNA for the synthesis of enzymes, 52

Genetics: the study of heredity, 2, 54

Genotype: the genic constitution of an individual, 54

Genus: a group of closely related species, 62

Geological eras: 100

Geotropism: a growth movement in response to gravity, 60

Germination: initiation of growth of a seed, bud, or spore, 46

Gibberellin: a plant hormone that causes stem elongation, 60

Ginkgo biloba: 92

Glucose: a simple sugar with the formula $C_6H_{12}O_6$, 6, 20, 22, 24, 40

Glycolysis: respiration of glucose to pyruvic acid, 22

Golden-brown algae: algae comprising the class Chrysophyceae of the division Chrysophyta, 68

Golgi apparatus: dictyosome; a cytoplasmic body believed to synthesize some carbohydrates, 10

Grain: a one-seeded indehiscent fruit in which the seed coat is fused at all points to the pericarp; caryopsis, 48

Grana: bodies within the chloroplast in which the light reactions of photosynthesis occur, 18

Grassland: a vegetation type in which the dominant vegetation consists of grasses, 104

Gravitational water: water that soil cannot hold against gravity and that drains off after a rain or flood, 34

Green algae: algae comprising the division Chlorophyta, 66, 68, 76, 84, 88, 100

Growth: increase in the number and size of cells in an organism, 2, 58

Growth movement: bending of an organ due to differential growth rates on opposite sides, 60

Growth ring: a xylem ring caused by changes in rate of cambial activity throughout the year, 40

Guard cell: one of a pair of cells surrounding a stoma and regulating its opening and closing, 14, 16, 32

Gullet: a groove in euglenoid cells from which flagella arise and through which food particles may be ingested, 66

Guttation: loss of water in liquid form from leaves, 30

Gymnospermae: the class of seed-bearing plants having ovules not enclosed in an ovary, 90, 100

Gynoecium: all the pistils of a flower collectively, 42

Haploid: having one set of chromosomes per nucleus, 42, 50, 62

Haustorium: in seed plants, a parasitic root, 28

Head: an inflorescence with many flowers tightly crowded on the flattened tip of the peduncle, 42

Heartwood: inner, nonfunctioning wood, 38

Hepaticae: a class of Bryophyta consisting of liverworts, 80

Heredity: the passage of genetic traits from parents to offspring, 54

Hesperidium: a berry with a leathery rind, 48

Heterogametes: gametes that are unlike in appearance, size, and behavior; eggs and sperms, 66

Heteromorphic: having sporophyte and gametophyte generations that are different in appearance, 62

Heterosporous: producing two types of sexual spores (megaspores and microspores), 84, 88, 90

Heterostyly: having styles of different lengths in different flowers, 96

Heterothallic: referring to gametophytes that produce gametes of only one sex or mating type and are not capable of self-fertilization, 62

Heterotrophic: referring to organisms that require an external supply of food, 70, 74, 100
Heterozygous: having two nonidentical genes of a pair, 54
Hilum: a scar remaining on the seed coat after the funiculus separates from the seed, 46
Holdfast: rootlike organ of attachment of some algae, 68
Homologous chromosomes: members of the same pair of chromosomes, 50
Homosporous: producing one type of sexual spore, 84, 88
Homothallic: referring to gametophytes producing gametes of opposite sex or mating type and capable of self-fertilization, 62
Homozygous: having two identical genes of a pair, 54
Hormone: a substance produced in one part of an organism and moving to another part where it exerts its effect; in plants, hormones control growth, differentiation, and tropisms, 58
Hornwort: nonvascular plants comprising the class Anthocerotae of the division Bryophyta, 78, 80
Horsetails: members of the subdivision Sphenopsida of the division Tracheophyta, 82, 100
Humus: partially decomposed organic material in soil, 34
Hybrid: an individual heterozygous for at least one pair of genes; the offspring of a cross between two species, 54, 98
Hybrid corn: high-yielding corn produced by crossing inbred strains, 56
Hybrid vigor: vigor characteristic of organisms heterozygous for many genes, 56
Hydathode: an opening at the margin of some leaves through which water of guttation exudes, 30
Hydrolysis: a chemical reaction in which a molecule is split into smaller molecules by the addition of water, 22
Hydrophyte: a plant adapted to living in water, 104
Hygroscopic water: water that is so closely bound to soil particles that it cannot be absorbed by plants, 34
Hymenium: fertile area in the fruiting body of some fungi, 74
Hypanthium: floral tube or calyx tube; the fused bases of sepals, petals, and stamens, 94
Hypha: a filament of a fungus, 74
Hypocotyl: the portion of a seed plant embryo below the cotyledons, 46, 90
Hypogynous: referring to flowers in which sepals, petals, and stamens arise from below the ovary, 94

IAA: indole-3-acetic acid, 58, 60
Imbibition: absorption of liquids or gasses into intermolecular spaces of a colloid, 32
Imperfect flower: a flower lacking either stamens or pistils, 44, 94
Imperfect fungi: fungi lacking sexual stages; Fungi Imperfecti; Deuteromycetes, 74
Impression: a fossil formed after a thin, buried plant part decays, 98
Inbreeding: crossing of closely related individuals, 56
Incomplete dominance: the contribution of two allelic genes to a phenotypic expression, 54
Incomplete flower: a flower lacking either sepals, petals, stamens, or pistils, 44, 94
Indehiscent: referring to fruits that do not split open at maturity, 48
Independent assortment, law of: Mendel's second law which states that at meiosis pairs of genes on different pairs of chromosomes segregate independently of each other, 54, 56
Indole-3-acetic acid: a naturally occurring auxin that functions in growth, tropisms, differentiation, and apical dominance, 58
Indusium: tissue covering the sorus of a fern, 86
Inferior ovary: ovary of an epigynous flower, 94
Inflorescence: a cluster of more or less closely associated flowers, 4, 42, 44, 94
Inheritance of acquired characteristics: 98
Insect pollination: 44, 96, 102
Integument: a layer of tissue surrounding an ovule, 42, 92
Interfascicular cambium: vascular cambium located between vascular bundles rather than within them, 40
Internode: region between two successive nodes on a stem, 4
Interphase: the phase between two successive nuclear divisions, 10, 12
Ion: an atom or group of atoms electrically charged by the gain or loss of electrons, 8
Irregular flower: bilaterally symmetrical flower, 94
Isogametes: gametes that are alike in appearance, size, and behavior, 66
Isomers: compounds having the same molecular formulas but different structural arrangements, 6
Isomorphic: having sporophyte and gametophyte generations that are alike in appearance, 62
Isotopes: atoms of the same element having the same atomic number but different atomic weights, 6

Karyogamy: fusion of two nuclei, 74
Key: samara, 48
Kinetochore: centromere; the region of a chromosome to which spindle fibers attach, 10
Kingdom: a group of divisions; the taxon of highest rank, 62
Kinin: a hormone that stimulates cell division, 60
Krebs cycle: the conversion of pyruvic acid to carbon dioxide by the stepwise removal of hydrogen, 22

Lamarck, Jean: a naturalist whose theory of evolution by inheritance of acquired characteristics is no longer accepted, 98
Laminarin: a stored food of the brown algae, 68
Lateral bud: a bud developing in the axil of a leaf, 4, 38, 60
Leaf: the primary photosynthetic organ of vascular plants, 2, 14, 82, 86, 90, 96
Leaf arrangements: 14
Leaf gap: an interruption in the stele caused by the departure of a leaf trace, 86
Leaf mosaic: a pattern formed by leaves exposing a maximum surface to the sun, 60
Leaf trace: a vascular strand connecting a leaf to the vascular tissue of the stem, 82
Leaflet: a segment of a compound leaf blade, 14
Legume: a dehiscent fruit that splits open along two opposite sides, 48
Lenticel: a structure in bark through which gaseous exchange occurs, 40
Lepidodendron: 84
Leucoplast: a colorless plastid in which food is stored, 10
Liana: a woody, climbing plant, 102
Lichen: a plant consisting of a fungus and an alga, 64, 76, 102

Light reactions: a series of light-requiring reactions in photosynthesis; ATP is synthesized and water is split into hydrogen and oxygen, 18
Linnaeus, Carl: an eighteenth-century naturalist whose system of naming organisms is used today, 64
Lip cells: thin-walled cells that interrupt the annulus of a fern, 86
Liverworts: nonvascular plants comprising the class Hepaticae of the division Bryophyta, 78, 80
Loam: a mixture of sand, silt, and clay, 34
Locule: a chamber in an ovary, 42
Long-day plant: a plant that blooms in response to long days and short nights, 58
Lycopodium: 82, 84
Lycopsida: a subdivision of cone-bearing, nonseed, vascular plants called club mosses in the division Tracheophyta, 82, 84, 100

Macromolecule: a large molecule, 8
Macronutrients: elements required by plants in relatively large amounts, 34
Maltose: a disaccharide, 6
Mannitol: a stored food in brown algae, 68
Marchantia: 78, 80
Margin: edge of a leaf, sepal, or petal, 14
Marginal placentation: placentation in which a single placenta extends the length of the ovary of a simple pistil, 96
Mass flow hypothesis: a hypothesis advanced to explain the movement of food in the phloem by differences in turgor pressure in leaves and roots, 40
Matter: anything that occupies space and has mass, 6
Megagametophyte: a gametophyte producing only female gametes, 82, 84, 90, 92, 94, 96
Megasporangium: a sporangium that produces megaspores, 84, 92, 94
Megaspore: a spore that can germinate into a female gametophyte, 42, 84, 92, 96
Megaspore mother cell: a diploid cell capable of producing megaspores by meiosis, 42, 92, 96
Megasporophyll: a sporophyll bearing one or more megasporangia, 82, 84, 90, 92; in flowering plants, a carpel, 94
Meiosis: a set of two nuclear divisions that reduce the number of chromosomes per nucleus by half and segregate homologous chromosomes and allelic genes, 42, 50, 52, 54, 62, 74, 86, 88, 92, 94
Meiospore: a spore produced by meiosis; a sexual spore, 62
Mendel, Gregor: an Austrian monk who discovered the laws of heredity, 54
Meristem: an embryonic tissue capable of cell division, 12
Mesocarp: the middle layer of the pericarp of a fruit, 46
Mesophyll: parenchyma tissue between the upper and lower epidermis of a leaf, 16
Mesophyte: a plant adapted to habitats that are neither very dry nor very wet, 104
Mesozoic era: 100
Messenger RNA: RNA that carries a copy of the genetic code from the DNA in the nucleus to the ribosomes in the cytoplasm, 50
Metabolism: the total of the chemical reactions occurring in an organism, 2
Metaphase: a phase of mitosis in which the duplicated chromosomes line up across the cell, 10, 12, 50
Microgametophyte: gametophyte producing only male gametes, 82, 84, 90, 92, 94
Micronutrients: elements required by plants in minute amounts; trace elements, 34
Microphyll: a leaf without an associated leaf gap and usually with a single unbranched vascular bundle, 82
Micropylar chamber: a chamber in a gymnosperm ovule in which the pollen grain remains after pollination and before fertilization, 92
Micropyle: an opening in the integument of an ovule through which the pollen tube usually enters; also the corresponding opening in the seed coat, 42, 46, 92
Microsporangium: a sporangium that produces microspores, 84, 92, 94
Microspore: a spore that can germinate into a male gametophyte, 44, 84, 92, 94
Microspore mother cell: a diploid cell capable of producing microspores by meiosis, 44, 92, 94
Microsporophyll: a sporophyll bearing one or more microsporangia, 82, 84, 92; in flowering plants, a stamen, 94
Middle lamella: cementing substance between adjacent cells, 10, 12
Midrib: main vein of a leaf, 14
Mineral nutrition: 34
Mitochondrion: a cytoplasmic body in which the Krebs cycle reactions and terminal oxidation occur, 10, 24
Mitosis: a nuclear division in which each daughter nucleus receives a set of chromosomes identical with that of the parent nucleus, 10, 12, 50, 74
Mold: a fossil formed when a bulky plant part buried in soil decays, 98
Molecule: smallest unit of a compound, 6
Monocaryotic: having one nucleus per cell, 76
Monocotyledonae: a subclass of angiosperms characterized by having one cotyledon in the seed, 94, 96
Monoecious: having male and female flowers (or cones) on the same plant, 44, 90
Monohybrid: heterozygous for one pair of genes, 56
Monosaccharide: a simple sugar, such as glucose, 6
Morphogenesis: the morphological and physiological changes that occur during the development of an organism, 58
Morphology: in botany, the study of structure and form of plants, 2
Mosses: nonvascular plants comprising the class Musci of the division Bryophyta, 80, 102
mRNA: messenger RNA, 50, 52
Multiple-accessory fruit: a multiple fruit that includes the floral tubes of the flowers and the peduncle on which they grew, 48
Multiple fruit: a fruit that develops from the pistils of several closely associated flowers, 48
Multiple gene inheritance: inheritance of traits influenced by several pairs of genes, 54
Musci: a class of Bryophyta consisting of mosses, 80
Mutation: a change in a gene, 52, 100
Mutualism: a relationship between two species in which both species benefit, sometimes called symbiosis, 102
Mycelium: all the hyphae forming the body of a fungus, 74
Mycology: the study of fungi, 2
Mycorhiza: a symbiotic combination of a fungus and a root,

26, 102; also such a relationship between a fungus and the subterranean gametophytes of some vascular plants, 82
Myxomycophyta: a division consisting of slime molds, 74

NAD: nicotinamide adenine dinucleotide, 22, 24
NADP: nicotinamide adenine dinucleotide phosphate, 18, 20
Nastic movement: response to an external stimulus; it is independent of the direction of the stimulus, 60
Natural selection: in Darwinian theory, the evolution of new species as a result of the survival of those individuals best adapted to the environment, 100
Neck: elongated upper portion of an archegonium through which a sperm reaches the egg, 78
Necrosis: death of certain parts of a plant, 34
Nectary: a gland secreting nectar, 44
Net venation: a network of branching veins in leaves, 14, 96
Neutron: an uncharged particle in the nucleus of an atom, 6
Nicotinamide adenine dinucleotide: NAD; a hydrogen carrier in cells, also called diphosphopyridine nucleotide (DPN), 22
Nicotinamide adenine dinucleotide phosphate: NADP; a hydrogen carrier in cells, also called triphosphopyridine nucleotide (TPN), 18
Nitrogen cycle: the cycle of nitrogen compounds within and among organisms and their environment, 72
Nitrogen fixation: conversion of free nitrogen of the air to a usable organic form, carried out by certain soil bacteria, 26, 72, 102
Node: the part of a stem at which leaves are borne, 4
Nodule: a swelling on the root of a leguminous plant in which nitrogen-fixing bacteria live, 26
Noncyclic photophosphorylation: one of the light reactions in photosynthesis; synthesis of ATP and the concomitant splitting of water into hydrogen and oxygen using light energy, 18
Nucellus: wall of the megasporangium of a seed plant, 42, 92, 94
Nuclear membrane: membrane surrounding a nucleus, 10
Nucleic acids: macromolecules that carry the genetic code (DNA) and control protein synthesis (RNA), 8, 12, 50
Nucleolus: a spherical, RNA-rich body in the nucleus, 10, 12
Nucleus: in the cell, a membrane-bound body containing the chromosomes, 10; in the atom, the central region containing protons and neutrons, 6
Nut: an indehiscent fruit resembling an achene but with a hard pericarp, 48
Nutation: spiral movement of a growing stem tip, 60

Oogonium: unicellular female reproductive organ of some members of the subkingdom Thallophyta, 66, 74
Oospore: a resistant spore formed by the fusion of an egg and a sperm, 66
Operculum: lid of a moss capsule, 80
Opposite leaf arrangement: arrangement of leaves in which two leaves occur at a node, 38
Order: a group of closely related families, 62
Organic compound: compound containing carbon, 8, 12, 100
Osmosis: diffusion through a differentially permeable membrane, 30
Osmotic pressure: potential pressure of a solution, 32
Outbreeding: crossing of unrelated individuals, 56
Ovary: the lower portion of a pistil containing one or more ovules, 42, 94
Ovulate cone: female cone of a gymnosperm, 90
Ovule: in a seed plant, a megasporangium surrounded by its integument(s), 42, 92, 94
Ovuliferous scale: ovule-bearing scale in the female cone of the Coniferales, 92
Oxidation: the loss of electrons (or hydrogen) from a substance, 18, 22
Oxidative phosphorylation: synthesis of ATP with the energy released by aerobic oxidation of glucose, 22

Paleobotany: the study of plant fossils and the evolution of plants, 2
Paleozoic era: 100
Palisade parenchyma: the primary photosynthetic tissue of a leaf; its cells are vertically elongated and usually comprise the upper part of the mesophyll, 14, 16
Palmate venation: vein pattern in leaves in which the main veins radiate from the tip of the petiole, 14
Pampas: grasslands of South America, 104
Panicle: a branched, racemous inflorescence, 42
Parallel venation: vein pattern in leaves in which the main veins lie approximately parallel to each other, 14, 96
Paramylum: a stored food of the Euglenophyta, 66
Paraphysis: sterile hair associated with reproductive organs of some fungi and algae, 80
Parasite: an organism that obtains food from a living host, 64, 70, 76, 104
Paratonic movement: growth movement controlled by external stimuli, 60
Parenchyma: unspecialized tissue composed of thin-walled cells, 12
Parietal placentation: placentation in which the ovules are attached to the outer wall of the ovary, 96
Passive absorption: absorption of water that is due to transpirational pull, 32
Pedicel: a stem bearing an individual flower of an inflorescence, 42
Peduncle: a stem bearing an isolated flower or inflorescence, 42
Penicillium: 76
Pepo: a berry with a hard rind, 48
Perfect flower: a flower possessing both stamens and pistils, 44, 94
Perianth: the calyx (sepals) and corolla (petals) of a flower, 42
Pericarp: fruit wall; ripened ovary wall, 46
Pericycle: a tissue lying between the vascular tissue and the endodermis; it gives rise to branch roots and to cork cambium, 26, 28
Perigynous; referring to flowers in which the sepals, petals, and stamens arise from a cup-shaped floral tube that surrounds the ovary but is not fused with it, 94
Periplast: rigid outer layer of cytoplasm found in some algae that lack cell walls (euglenoids), 66
Perisperm: food-storage tissue that develops from the nucellus in a few species of flowering plants, 46
Peristome teeth: hygroscopic teeth that surround the mouth of a moss capsule and aid in spore dissemination, 80
Perithecium: the flask-shaped ascocarp of some Ascomycetes, 76
Permanent wilting percentage: the percentage of water left in

a soil when the plants growing in it have reached permanent wilting, 34
Petal: a member of a whorl of sterile appendages situated between the calyx and the fertile organs of a flower, 42, 94
Petiole: leaf stalk, 14
Petrifaction: a fossil in which the cell wall material has been replaced by minerals, 98
pH: a measure of acidity or alkalinity, 36
Phaeophyta: a division consisting of brown algae, 68
Phellem: cork, 28
Phelloderm: a parenchymalike tissue produced by cork cambium in some species of vascular plants, 26, 28
Phellogen: cork cambium, 28
Phenotype: the appearance or physiological characteristics of an individual; the visible manifestation of the genotype, 54
Phloem: food-conducting tissue of vascular plants, 12, 16, 28, 38, 40, 90
Photoperiodism: response of plants to day length, 58
Photosynthesis: manufacture of food from carbon dioxide and water, using light energy, 18, 32, 72, 100
Phototropism: a growth movement in response to unilateral presentation of light, 60
Phycocyanin: a blue photosynthetic pigment found in the blue-green algae and the red algae, 66
Phycoerythrin: a red photosynthetic pigment found in the red algae and the blue-green algae, 66, 68
Phycology: the study of algae, 2
Phycomycetes: a class consisting of algalike fungi, 74
Phylum: a group of related classes; division, 62
Physiology: the study of functions of cells, tissues, or organs, 2
Phytochrome: a photoreceptive pigment that functions in photoperiodic responses, 58
Pinna: leaflet of a compound fern leaf, 86
Pinnate venation: vein pattern in leaves in which secondary veins branch off from the midrib at regular intervals, 14
Pioneer plants: first plants to colonize newly exposed land, 80
Pistil: the female organ of a flower; it consists of one or more carpels, 42, 94, 96
Pith: the central tissue of stems and of monocot roots, 28, 38
Placenta: the portion of an ovary to which ovules are attached, 42, 94
Placentation: the arrangement of placentas in the ovary, 94
Plankton: aquatic organisms that float or are unable to swim against currents, 68
Plasma membrane: outer membrane bounding the cytoplasm of a cell, 10
Plasmodesmata: thin strands of cytoplasm extending through the cell walls of adjacent cells, 10
Plasmodium: a multinucleate mass of protoplasm without cell walls; the vegetative stage in the life cycle of a slime mold, 74
Plasmogamy: fusion of protoplasts of two gametes without fusion of their nuclei, 74
Plasmolysis: shrinking of cytoplasm from the cell wall due to movement of water from the protoplast into a solution with lower diffusion pressure, 30
Plastid: a cytoplasmic body that usually manufactures food or stores it, 10
Plumule: epicotyl, 46, 90
Polar nuclei: the two nuclei of an endosperm mother cell which when fertilized by a sperm nucleus become the polyploid endosperm nucleus, 44, 96
Polarity: morphological or physiological differentiation of opposite ends of an organ or a plant, 60
Pollen chamber: in angiosperms, the chamber in the anther in which pollen grains are produced, 44; in gymnosperms, the micropylar chamber, 92, 94
Pollen grain: the microgametophyte of a seed plant, 42, 44, 90, 92, 94
Pollen mother cell: microspore mother cell of a seed plant, 44, 94
Pollen sac: microsporangium of seed plants, 44, 92, 94
Pollen tube: a tube formed by a pollen grain; it grows toward the female gametophyte and discharges its sperms into it, 44, 90, 92, 94
Pollination: transfer of pollen grains from a stamen to a stigma (in angiosperms) or from a staminate cone to an ovule (in gymnosperms), 44, 96
Polyploid: having more than two sets of chromosomes per nucleus, 52, 98
Polysaccharide: macromolecule composed of repeating monosaccharide subunits, 6
Pome: a fruit resembling a berry but with most of the outer fleshy portion derived from floral parts, 48
Prairie: grassland of North America, 104
Precambrian era: 100
Primary cell wall: a thin cellulose wall possessed by most plant cells, 10
Primary leaf: first leaf produced by an embryo, 88
Primary root: first root produced by an embryo, 26, 46, 88
Primary succession: succession on newly exposed land, 102
Primary tissue: tissue produced by a primary (apical) meristem, 28, 38
Proembryo: the first group of cells formed from the zygote; the embryo is derived from certain of these cells, 46, 90
Prop root: adventitious roots that usually arise at a point on a stem above the ground, 28
Prophase: an early stage of mitosis in which the chromosomes shorten and thicken and are first visibly double, 10, 12, 50
Protein: a macromolecule composed of amino acid subunits, 8, 22
Proterozoic era: 100
Prothallus: gametophyte of a fern, psilophyte, club moss, or horsetail, 82, 86, 88
Proton: positively charged particle in the nucleus of an atom, 6
Protonema: the early, filamentous stage in the development of a bryophyte gametophyte, 80
Protoplast: the living portion of cells or organisms, 10
Protostele: a solid core of primary vascular tissue, 86
Psilopsida: a subdivision of rootless, nonseed, vascular plants in the division Tracheophyta, 82, 84, 100
Psilotum: 82, 84
Pteridospermales: seed ferns, an order of extinct gymnosperms, 92
Pteropsida: a subdivision of the division Tracheophyta; it includes ferns and seed plants, 82, 86, 100
Puccinia graminis: 76
Purine: a nitrogen base present in nucleic acids and some coenzymes, 50
Pycnium: spermatogonium of wheat rust, 76
Pyrenoid: a starch-storing body in the chloroplasts of some algae and hornworts, 66, 78
Pyrimidine: a nitrogen base present in nucleic acids, 50

Pyrrophyta: a division of algae including dinoflagellates and cryptomonads, 66, 68

Raceme: an inflorescence with a long peduncle from which branch the pedicels of the flowers; the pedicels are the same length, 42
Rachis: in pinnately compound leaves, the extension of the petiole from which the leaflets arise, 14
Radicle: the meristematic portion of a seed plant embryo that produces the primary root, 46, 90
Radiocarbon dating: a method of calculating the age of fossils by determining their radioactive carbon content, 98
Raphe: a ridge formed by the fusion of the funiculus with the seed coat, 46; a slit in the cell wall of diatoms, 68
Rays: horizontal conducting tissue in xylem and phloem, 38, 90
Receptacle: the tip of a flower-bearing stem, 42, 94
Recessive gene: a gene that does not produce a phenotypic effect in the presence of its dominant allele, 54
Red algae: algae that comprise the division Rhodophyta, 68
Reduction: the addition of electrons (or hydrogen) to a substance, 18
Region of elongation: a region of roots and stems in which cells increase in length, 26, 38
Region of maturation: a region of roots and stems in which cells differentiate into mature tissue, 26, 38
Regular flower: radially symmetrical flower, 94
Replication: exact duplication of genetic material, 52
Reproduction: production of offspring by sexual or asexual means, 2
Resin: a complex mixture of compounds produced in the wood of some gymnosperms and other trees, 90
Respiration: oxidation of food with the release of energy that can be used in synthetic reactions, 22, 100
Rhizobium: 26, 72
Rhizoid: elongated, rootlike, absorbing cell or filament of cells in some Thallophyta and in gametophytes of some Embryophyta, 78, 80
Rhizome: horizontal underground stem, 40, 86, 88
Rhizophore: the root-bearing organ of *Selaginella*, 82
Rhizopus: 74
Rhodophyta: a division comprising the red algae, 68
Rhynia: 84, 100
Ribonucleic acid: nucleic acid involved in protein synthesis, 50
Ribosomal RNA: RNA of ribosomes on which protein is synthesized, 50, 52
Ribosome: a small cytoplasmic body that is the site of protein synthesis in the cell, 10, 52
Ricciocarpus: 78, 80
Ring porous: referring to wood with larger or more numerous vessels in the spring wood than in the summer wood, 38
RNA: ribonucleic acid, 50
Root: a cylindrical, usually branched, vascular organ that anchors the plant and absorbs water and minerals, 2, 26, 30
Root cap: a mass of cells that sheath the apical meristem of the root and protect it from mechanical injury, 26
Root hair: a fine extension of a root epidermal cell that increases the water-absorbing capacity of the root, 26, 30
Root hair zone: region of a root that bears root hairs, 26
Root pressure: pressure developed within a root when the rate of water absorption exceeds that of transpiration, 30, 32
Rootstock: rhizome, 40
Runner: a horizontal stem growing on the surface of the ground, 40
Rust: a basidiomycete that parasitizes cereal grains, 76

Sac fungi: fungi comprising the class Ascomycetes and characterized by the production of sexual spores borne in asci, 74
Samara: an indehiscent fruit resembling an achene but bearing a wing, 48
Sand: a soil with particles 0.02 to 2.00 mm. in diameter, 34
Saprolegnia: 74
Saprophyte: an organism that obtains its food from dead organisms or from the waste products of living organisms, 64, 66, 70, 104
Sapwood: outer, functional wood, 38
Sarcina: a cubical group of eight bacterial cells, 70
Savannah: grassland with scattered trees, 104
Scarification: scratching or breaking of seed coats by abrasion or freezing and thawing, 48
Schizocarp: a dry fruit that splits into achene-like segments, 48
Schizomycophyta: a division consisting of bacteria, 70
Scientific method: 4
Sclerenchyma: a strengthening tissue of dead, thick-walled cells, 12
Scutellum: cotyledon of a grass embryo, 46
Secondary cell wall: a thick, inner cell wall formed after the primary wall, 10
Secondary root: branch root, 26
Secondary succession: succession on disturbed land, 102
Secondary tissue: tissue produced by secondary (lateral) meristems; secondary xylem, secondary phloem, cork, or phelloderm, 26, 28, 40
Seed: a ripened ovule consisting of a seed coat, an embryo, and, in some species, endosperm or perisperm, 4, 44, 46, 90, 92, 94, 96
Seed coat: the ripened integument surrounding a seed, 46, 92
Seed ferns: extinct gymnosperms comprising the order Pteridospermales, 92, 100
Seed plants: gymnosperms and angiosperms, 90
Segregation, law of: Mendel's first law which states that two alleles of a pair segregate from each other during meiosis and pass into different gametes, 54, 56
Selaginella: 82, 84
Self-pollination: pollination of a pistil by stamens from the same plant, 44, 54, 96
Sepal: one of the outermost whorl of sterile appendages of the flower, 42, 94
Sessile: referring to a leaf without a petiole or a flower without a pedicel, 14, 42
Seta: stalk of a moss sporophyte, 78
Sexual reproduction: reproduction that involves both fusion of gametes and meiosis, 42, 66
Sexual spore: a spore that is the product of meiosis, 62, 76
Shoot: a young stem and its leaves, 4
Short-day plant: a plant that blooms in response to short days and long nights, 58
Sieve plate: pitted end wall of a sieve-tube element, 12, 38
Sieve tube: a food-conducting phloem tube composed of sieve-tube elements, 12, 38
Sieve-tube element: a food-conducting phloem cell that lacks a nucleus at maturity, 38
Sigillaria: 84

Silt: a soil with particles 0.002 to 0.02 mm. in diameter, 34

Siphonostele: a ring of primary vascular tissue around a central column of pith, 86

Slime layer: capsule surrounding a bacterial cell, 70

Slime molds: fungi comprising the phylum Myxomycophyta and having some animal-like characteristics, 74

Smut: a basidiomycete that parasitizes cereal grains, 76

Soil: a mixture of mineral and organic particles that covers most of the land surface of the earth, 34

Solenostele: a siphonostele with only one leaf gap at any one level, 86

Sorus: a cluster of sporangia on a fern leaf, 86

Spacial isolation: physical separation preventing interbreeding of two or more groups of organisms, 98

Species: a group of organisms that interbreed freely, 62, 98

Sperm: motile male gamete, 44, 66

Spermagonium: stage in the life cycle of wheat rust, 76

Spermatangium: male reproductive organ of most red algae, 68

Spermatium: nonmotile male gamete, 68, 76

Sphagnum: 80

Sphenopsida: a subdivision of cone-bearing, nonseed, vascular plants called horsetails in the division Tracheophyta, 82, 84, 100

Spike: an inflorescence with a long peduncle bearing sessile flowers, 42

Spindle fiber: an extremely fine fiber that extends from one pole of a dividing cell to the centromere of a chromatid or from pole to pole, 10

Spine: a leaf or leaf part modified into a sharply pointed structure, 16

Spirillum: a spiral-shaped bacterial cell, 70

Spongy parenchyma: a tissue, usually in the lower part of the mesophyll of a leaf, composed of irregularly shaped cells with large intercellular spaces, 14, 16

Sporangium: a structure within which spores are produced, 74, 82, 84, 86

Spore: a single cell other than a zygote or a gamete capable of giving rise to an entire plant, 62, 70

Sporophyll: a leaf bearing one or more sporangia, 84, 86

Sporophyte: a plant that produces sexual spores; the diploid generation, 62, 68, 78, 80, 82, 84, 86, 90, 94

Stalk cell: a vegetative cell in a gymnosperm pollen grain, 90

Stamen: the microsporophyll of a flowering plant; it consists of an anther, in which pollen is produced, and a filament, 42, 94

Staminate cone: male cone of a gymnosperm, 92

Staphylococcus: bacterium having spherical cells occurring in pairs, 70

Starch: a common form of food reserve in plants; a long-chain polysaccharide composed of glucose subunits, 6, 12, 22, 66, 78

Stele: central cylinder of a stem or root, 28, 86

Stem: a cylindrical, usually branched, vascular organ that supports leaves and flowers (or cones), 2, 38, 90

Stem modifications: 40

Steppe: grassland of Asia, 104

Stigma: uppermost part of the pistil that is receptive to pollen, 42, 94

Stipules: two appendages that occur at the base of the petioles of some leaves, 14

Stolon: in vascular plants, a horizontal stem growing along the surface of the ground, 40

Stoma (plural, **stomata**): opening in the epidermis of a leaf or stem through which gases diffuse, 14, 16, 32, 38, 82

Storage capacity: the field capacity of a soil minus its permanent wilting percentage, 34

Streptococcus: bacterium having spherical cells occurring in chains, 70

Strobilus: a grouping of sporophylls; a cone, 84

Struggle for existence: competition between individuals for food or living space, 100

Style: the portion of the pistil between the stigma and the ovary, 42, 94

Succession: a series of communities that sequentially occupy exposed or disturbed land, 102

Succulent: a plant, often fleshy, with pronounced water-storing tissues, 16, 40, 104

Sucrose: a disaccharide, 6, 40

Superior ovary: ovary of a hypogynous or perigynous flower, 94

Survival of the fittest: survival and reproduction of individuals best adapted to the environment, 100

Suspensor: in vascular plants, a portion of the embryo that pushes the embryo proper into the gametophyte, 46, 90; in some Phycomycetes, one of the short hyphae supporting the zygospore, 74

Symbiosis: a relationship between two species of organism; sometimes used synonymously with mutualism, sometimes used to include parasitism, 26, 64

Synapsis: pairing of homologous chromosomes during prophase I of meiosis, 50

Synergid cells: cells associated with the egg in the embryo sac of angiosperms, 42, 44

Tap root system: a root system with a thick primary root and smaller branch roots, 26

Tapetum: nutritive tissue in the sporangia of many tracheophytes, 44, 82

Taxon: a unit of classification, such as species, genus, and family, 62

Taxonomy: the science of naming, classifying, and identifying organisms, 2, 62

Telium: a stage in the life cycle of wheat rust, 76

Telophase: a late stage of mitosis in which the chromosomes reorganize into two daughter nuclei, 12, 50

Temperate deciduous forest: 102

Tendril: a leaf or stem modified into a twining organ, 16, 40

Tepal: a sterile appendage of a perianth not differentiated into sepals and petals, 42

Terminal bud: a bud developing at the tip of a stem, 4, 38

Terminal oxidation: in respiration, the transfer of hydrogen removed from a food molecule to molecular oxygen with the formation of water and the release of energy, 22, 24

Test cross: a genetic test to determine homozygosity or heterozygosity of an individual, 56

Thallophyta: the subkingdom composed of plants that do not retain embryos within the female reproductive organ; the plant body is a thallus, and reproductive organs are unicellular; includes the algae, bacteria, and fungi, 64

Thallus: a plant body without true roots, stems, or leaves, 66

Thigmotropism: a growth movement in response to touch, 60

Thorn: a stem modified into a woody, pointed structure, 40

Tissue: a group of similar cells that perform the same function, 12, 16

TPN: triphosphopyridine nucleotide, 18
Trace element: an element required in minute amounts; a micronutrient, 34
Tracheid: an elongated conducting and supporting xylem cell, 12, 38, 86, 90
Tracheophyta: a division of the Embryophyta including all vascular plants, 64, 82, 100
Transfer RNA: RNA that assembles amino acids on a ribosome in a sequence in accord with the genetic code carried by messenger RNA, 50
Translocation: in plants, the movement of food and water from one part to another, 40; in genetics, the exchange of broken fragments between nonhomologous chromosomes, 52
Transpiration: loss of water in vapor form from plants, 14, 30, 32, 82
Trichogyne: an extension of the ascogonium of Ascomycetes, 76; an extension of the carpogonium of red algae, 68
Triphosphopyridine nucleotide: TPN; a hydrogen carrier in cells; also called nicotinamide adenine dinucleotide phosphate (NADP), 18
Triple fusion nucleus: endosperm nucleus formed by the fusion of two polar nuclei and a sperm nucleus, 44, 96
Triplet code: the genetic code, a code in which three nucleotides code for one amino acid, 52
tRNA: transfer RNA, 50, 52
Tropical deciduous forest: 104
Tropical rain forest: 104
Tropism: an induced growth movement in response to a unilateral stimulus, 60
Tube cell: the cell of a pollen grain that produces the pollen tube, 44, 90, 94
Tuber: enlarged, fleshy portion of a rhizome, 40
Tundra: a vegetation type characteristic of areas with permanently frozen subsoil; mosses and lichens predominate, 102, 104
Turgor pressure: pressure within a cell due to absorption of water, 30, 32

Umbel: an inflorescence in which the pedicels of all the flowers arise from the tip of the peduncle, 42
Uredinium: a stage in the life cycle of wheat rust, 76

Vacuolar membrane: membrane bounding a vacuole, 12
Vacuole: a cavity within the cytoplasm containing dissolved substances, 12, 30
Valence: the number of electrons an atom tends to gain, lose, or share, 6
Variety: a subdivision of a species, 62
Vascular bundle: a strand of primary xylem and primary phloem, 14, 16, 38, 40, 96
Vascular cambium: a secondary (lateral) meristem that produces secondary xylem and secondary phloem, 26, 28, 40
Vein: a vascular bundle in a leaf, 14
Velamen: the spongy epidermis of some aerial roots that enables them to absorb water quickly, 28
Venation: arrangement of veins in a leaf, 14
Venter: the basal, egg-containing portion of an archegonium, 78, 82, 88
Ventral canal cell: a sterile cell associated with an egg in the venter of an archegonium, 78, 90
Vessel: a water-conducting xylem tube composed of vessel elements, 12, 38, 94
Vessel element: a water-conducting xylem cell that lacks end walls, 38
Viruses: submicroscopic, acellular, infectious agents that have only some of the characteristics of living things, 72
Vitamins: organic compounds that are required in minute amounts and have enzymatic functions, 58

Wheat rust: a basidiomycete parasite of wheat, 76
Whorled leaf arrangement: arrangement of leaves in which three or more leaves occur at a node, 38
Wind-pollination: 44, 92, 96
Wood: secondary xylem, 38, 40

Xanthophyceae: yellow-green algae, 68
Xanthophyll: a yellow pigment associated with chlorophyll, 16, 66, 68, 78
Xerophyte: a plant adapted to growing in very dry habitats, 104
Xylem: water-conducting tissue of vascular plants, 12, 16, 28, 30, 38, 40, 82, 86, 90

Yeasts: single-celled Ascomycetes without fruiting bodies, 76
Yellow-green algae: algae comprising the class Xanthophyceae of the division Chrysophyta, 68

Zoospore: a motile spore, 66
Zygomorphic: referring to a bilaterally symmetrical flower, 94
Zygospore: a resistant spore formed by the fusion of isogametes, 66, 74
Zygote: cell resulting from the fusion of two gametes; the first cell of a sexually produced individual, 44, 62, 66